U0396197

来自郭吴的消息

News from Somewhere Called Zhangwu

十二楼建筑工作室作品集

Selected Works of Architect Studio 12F

贺　勇　等 编著

Written by He Yong

东南大学出版社 · 南京

SOUTHEAST UNIVERSITY PRESS · NANJING

内容提要

本书是十二楼建筑工作室的作品集，收录了近十年来该工作室扎根在浙江省安吉县鄣吴镇完成的十余项建筑作品，通过对每个建筑的分析、解读以及其背后故事的讲述，探讨场地、生活以及空间之间的关系，阐述了其面向土地与日常生活的设计理念与工作方式。

图书在版编目（CIP）数据

来自鄣吴的消息：十二楼建筑工作室作品集 / 贺勇等编著．—南京：东南大学出版社，2019.11
ISBN 978-7-5641-8535-0

Ⅰ．①来… Ⅱ．①贺… Ⅲ．①建筑设计－作品集－中国－现代 Ⅳ．①TU 206

中国版本图书馆 CIP 数据核字（2019）第 188459 号

来自鄣吴的消息：十二楼建筑工作室作品集
Laizi Zhangwu De Xiaoxi：Shierlou Jianzhu Gongzuoshi Zuopinji

编　　著：贺　勇　等
责任编辑：宋华莉（52145104@qq.com）

出版发行：东南大学出版社
社　　址：南京市四牌楼 2 号
邮　　编：210096
网　　址：http://www.seupress.com
出 版 人：江建中

印　　刷：上海雅昌艺术印刷有限公司
开　　本：889 mm × 1194 mm　1 / 24　印张：12.25　字数：546 千字
版 印 次：2019 年 11 月第 1 版　2019 年 11 月第 1 次印刷
书　　号：ISBN 978-7-5641-8535-0
定　　价：118.00 元

经　　销：全国各地新华书店
发行热线：025-83790519　83791830

目录 Contents

前言 Foreword

鄣吴镇，位于浙江省北部的安吉县，距离杭州约一个半小时的车程。全镇人口1.2万，分布在鄣吴村、玉华村、景坞村等6个行政村里，其中鄣吴村规模最大、人口最多，也是镇政府所在地。鄣吴村保留下来的老房子很少，绝大多数房子都是最近30年建造的，但是其历史上由“八府九弄十二巷”以及穿村小溪构成的空间格局却相对完整地保留了下来，建筑也维持着高密度的紧凑状态，所以该村依然显示出一种比较强烈的传统风貌。

自2010年，十二楼建筑工作室开始介入鄣吴镇各村落的规划与建设，陆续完成了数十项公共建筑与设施的规划与设计。本书中收录的所有建筑就是我们在鄣吴镇工作的成果。这些建筑改变了那里的乡村，也改变了我对建筑实践与教学，乃至生活的态度。在乡村的工作和生活让我似乎重新建立了感知世界的通道与桥梁，让我时常静下心来反思我们在城市里早已固化甚至有点异化的建筑教育以及设计实践。

相比深受政治、经济、意识形态的影响而“变异”的城市建筑，乡村建筑更加真实地反映了居民的生产、生活与其所依存的土地之间的关系，因而显得质朴、自然。在这样一种背景之下，乡村建筑，在当下作为一种机制与方法，其意义远胜过作为一种固有风貌或审美对象，因为它具有更加普适性的价值。所以在我的研究生教学中，我都带着他们进行一定数量的村庄调研，并寻找在乡村中建成一个小房子的机会。从场地的踏勘、测量开始，聆听村民的意见、寻找适宜的建造材料、反复画图与修改、破土动工、与工匠师傅交流、控制造价，让他们感受设计、建造的每一个环节，以此来理解设计师的角色与任务。设计的目标是满足人的使用需求，而不单是表达建筑师的个人意愿；设计是理解、倾听、包容的过程；解决问题的手段绝不仅是概念、空间、形式，还包括建造方式、运营模式；建筑无论大小，都要注入情感，需要一种诗性与智慧……或许，这才是一个建筑师应有的工作态度与方法。在这个过程中，建筑师往往用一种引导而非主导的工作方式，所以其建成结果可能与最初的设计相差甚远。以本作品集中的玉华村竹酒设施用房为例，最后建成房子的建筑师是我们还是业主老陈？这着实难以分辨，但是又有什么关系呢，因为重要的是那些房子融入了场地，满足了功能的需求，并受到了业主和游客的喜爱。

随着时代的发展，各种知识可以轻而易举地收集与获得，人们似乎根本无须亲历现场；面对一切事情，我们开始习惯性地参照他人的点评与攻略，不知不觉中，我们头脑中已经有了太多“经验”与“程式”。时间一长，很多学生、建筑师都会觉得不同的设计都大同小异，逐渐丧失设计的热情与初衷。面对这一困境，现代建筑的先驱以及教育大家格罗皮乌斯的一段话在今天或许依然具有很强的启示，他说学习设计最要紧的是保持一种“没有被理性知识的积累所影响的新鲜心灵”。如何才能真正做到呢？或许适度抛弃那些既定的经验与程式，回到鲜活的土地与丰富的生活是一种可能且有效的方法。如果说我们在鄣吴的坚守与实践传递了一些消息，我想这应该就是其中最有价值的部分吧。

最后，要感谢曾经参与过这些项目的所有老师和同学，以及帮助这些项目得以落地建成的地方领导、工匠师傅；感谢尹子良、陈钰凡同学的精心排版；感谢郝军同学贡献了许多精彩的照片；感谢侯艳平同学高质量的英文翻译。

贺勇 He Yong
2019年1月于杭州
January, 2019, in Hangzhou

The town of Zhangwu is located in Anji County, northern part of Zhejiang Province. It only takes one and a half hours to get there from Hangzhou by driving. The town has a population of 12,000. It consists of six villages, among which Zhangwu village occupies the biggest area with the largest population. It is also the site of the town government. The whole village is fairly new since most of the houses there were built within the last 30 years. And yet it boasts distinct Chinese traditional styles and features since its old layout, known as "eight mansions, nine lanes and twelve alleys" as well as a rivulet traversing the whole village has been well preserved.

Since the year 2010, our studio has been deeply involved in the planning and construction of all the villages in Zhangwu Town. In fact, we have successively completed the designing and planning for over a dozen of public facilities and buildings. All the buildings introduced and discussed in this book are the results of our work in Zhangwu Town. They have not only changed the landscape of those villages but also my own philosophy of architecture work and teaching. What's more, they have even changed my attitude toward life itself. My work and life in rural villages seem to have opened a new window for me to perceive the world, which compels me to really calm down and reflect on the somewhat stereotypical, even mutated architectural work and education in urban areas.

Compared with the "mutated" urban architecture, rural buildings look more rustic and natural because they reflect, to a larger extent, real life and work of the local people as well as their connection to the land. And therefore, rural architecture as a mechanism and method has far more significance an inherent style or aesthetic objects— it has more universal value. That's why I would prefer to take my postgraduate students to several villages for close research, trying to find an opportunity for them to build a small house there. From the measuring of the site to listening to villagers' opinions, finding suitable materials, drawing and modifying the designing chart, and to controlling cost, communicating with construction workers, I let them get involved in every step of the whole procedure so that they could better understand an architect's job. The goal of design is to cater for the needs of users, not for the self-expression of architects. Design entails understanding, listening and toleration. The solutions to problems are by no means only concept, space, form. Instead they also include the practice of building strategies and operating models. Architecture, no matter how big or small it is needs to be injected in emotions and passion... All of those, perhaps are the right method and attitude that an architect should have. More often than not, an architect does not play the dominant role but merely a guiding one. That's why the ultimate result of a designing work might end up a lot different from its initial version. Taking the Bamboo-wine Brewery Building of Yuhua Village as an example, who finally built the building, we or the owner, Lao Chen? It's hard to tell. But why should that even matter? What's important is that the finished buildings get integrated into its environment easily, manage to serve their intended functions perfectly and are admired and adored by both their owners and visitors.

With the development of the society and the technology, oceans of knowledge and information are just a click away. It seems that there is no need to go through the scene at all. We have been used to doing everything based on other people's "reviews" and "strategies" which can be found online. Gradually, our minds are filled with "experience" and "formulated procedures". Over time, many architects and students of architecture are losing their passions and initial aspiration for architectural designing, thinking all architecture is the same. To deal with this problem, a quotation from Walter Gropius, the great educator and pioneer of modern architecture in early 20th century might be particularly inspirational even for today. He once said that the most important thing to do in the study of design is to strive to maintain a "fresh mind unaffected by the accumulation of rational knowledge". How can it be achieved? Perhaps the answer is to rid oneself of certain stereotypical procedures or experience and return to the land and life that are rich and alive. I think this should be the most valuable news sent from our practice in Zhangwu.

Finally, I would like to extend my gratitude to all my students and colleagues who have participated in my projects in Zhangwu. I would also like to thank the supports from local authorities and craftsmen, the meticulous typesetting by Yin Ziliang and Chen Yufan, the beautiful photos provided by Hao Jun as well as the faithful translation for this book by Hou Yanping.

鄣吴镇全景 The Panoramic View of Zhangwu Town

观点 1 让建筑回到土地，回到日常

Viewpoint 1 Bring Architecture to the Ground and Daily Life

贺勇，浙江大学建筑系，十二楼建筑工作室 I He Yong, Architecture Department of Zhejiang University, Architect Studio 12F

这些文字是2016年4月接受建筑媒体《有方》的访谈，在很大程度上反映了我这些年的心路历程及以及一些相关思考，故收录在此，以作为本作品集的部分观点。

1 最近在做的项目是哪些？

我的身份首先是大学教师，所以很大一部分时间自然是花在了教学上面。今年春、夏学期我带着四年级本科生，继续着这几年一直在进行的“面向土地与生活的建筑设计”的课程教学，探讨如何让建筑回到土地、回到节令、回到与生活的真实联系。课程目的旨在修正我们以往过于关注空间、形式的价值观念与工作模式，换一种方式来讨论建筑的可能。今年的项目是浙江安吉县鄣吴镇玉华村茶厂的改造，并有一个有趣的英文标题：TEA PLUS。基本背景是由于采茶、制茶的季节性，茶厂建筑在大部分时间是闲置的，于是我要求同学们探讨加入其他功能的必要性与可能性。4月初，我带着参加该课程的同学到茶厂调研，并真的在茶园里采茶半天，时间虽短，但在体验采茶、制茶人的劳作与生活状态之后，同学们真切感受到了生活中有诗歌，更有辛苦，甚至心酸。我相信他们在回校之后，对设计的目的与意义相比以前有了更多不一样的理解。

Below is part of the transcription of an interview I had with *You Fang* (Position in English), a Chinese architectural institution in April, 2016. I believe that my answers to these questions have reflected, to a great extent, the evolving of my feelings and thoughts about architecture over these recent years, hence its inclusion as the preface of this book.

1 Are there any particular projects you are working on recently?

I am, first and foremost, a university professor, so naturally, most of my time is dedicated to my teaching. My teaching mainly focuses on "architectural designs for life and land", which is an ongoing theme of exploring the ways to bring architecture back in close contact with the land, seasonal changes and daily life. My course aims to rectify the old designing philosophy which over-emphasizes space and form, and to discuss the possibility of an alternative approach to architectural designing. For this spring-summer term, I have a class of senior year undergraduates to teach, a big responsibility. My undergraduate fourth-grade students and I continued to explore the above theme and completed a tea factory renovation project. The project is located at Yuhua village in Zhangwu Town, Anji County, Zhejiang Province. I gave the project an English title: TEA PLUS, which I find quite appealing. The project is based on the fact that the tea factory is out of use and hence vacant most of the year due to the seasonal characteristic of tea-processing business. Therefore, I encouraged my students to

作为一个建筑师，我现在做的设计项目主要是在鄣吴镇，2010年以来，我们在那里完成了十余项小房子设计项目，主要是各类公共性的服务设施，例如卫生院、乡村社区中心、小卖店、公交站、垃圾站、公厕等。最近正在进行中的是该镇玉华村、景坞村的村委会与社区中心，接下来可能还要设计一个乡村公墓。这种基于村民日常生活真实需求的设计与建造，总是让我的内心感到欢喜与安定，因为我看到这些房子确实改善了当地居民的生活，让我确信这是一件有意义的事情。在我眼中，有人用、愿意用的房子才是真正的好房子。

2 和过往比，最近做的项目有哪些新的思考或尝试?

10年以前，我和身边的大多数朋友一样，受委托或投标，设计着天南海北、大大小小的各类建筑，也陆续建成了一些，但没有一个让自己满意的。尽管那些项目所提出的概念往往颇具匠心，图纸上看起来也多美轮美奂，但建成后的情形每次都着实让人失望，甚至迷茫。2007年在慕尼黑工业大学学习，让我看到了很多欧洲建筑师的工作状态，特别是德国、瑞士建筑师。他们中的很多工作于一些很小的城镇或乡村，事务所多是三五个人的小规模，安安静静设计着当地的小房子，如此方式，使他们得以深入了解当地的建造知识与经验，与建设的各个方面保持密集而有效的沟通，并频繁地前往工地进行监督、指导。对他们而言，高品质的设计与建造是一件普遍、自然的事情。于是从那时开始，“如何做当地人，设计自己可控的小房子”成为我心底的渴望与目标，也缘于各种机缘巧合，有了现在位于鄣吴镇的系列实践与教学，这些事情刚开始的时候只是一时兴趣，可是最近越来越感觉到那些琐碎的、片段式的东西开始逐步整合连贯起来，构成了某种相对完整的方法，那就是面向土地与生活的建筑教育与设计实践，让建筑回到土地，回到日常。鄣吴镇这片土地，很大程度上改变了我以往对于城市建筑的实践与教学，乃至生活的态度。在我现在的

have discussions about whether it is necessary to add some extra functions to the existing facilities and what functions it might be. At the beginning of this April, I took them to the factory for a research, during which we spent half a day picking tea-leaves side by side with local workers. It was a short yet inspirational experience. My students learnt a valuable lesson about the hardships, as well as the beauty of manual labor. I believe they gained a rather different understanding about the purpose and meanings of design than before.

As an architect, I mainly focus on the projects in Zhangwu Town. Starting from the year 2010, my associates and I have completed dozens of designing works on small buildings which are mainly public facilities such as a village hospital, a village community center, a convenience store, a bus station, a garbage disposal station, and several public toilets, to name just a few, etc. Recent ongoing projects include the village committees and community centers in Yuhua village and Jingwu village, and possibly a village cemetery. Projects of designing and construction catering to the needs of local villagers in their daily life such as these make me believe that what I am doing actually means something meaningful. For me, the only good buildings are those really useful to and being used by people.

2 Compared with your previous projects, is there anything new that you are trying out on your recent ones?

10 years ago, like most of my friends and colleagues, I took on any projects that I could get my hands on without discrimination, buildings of various sizes and scales at locations far across the world. My works were plenty but few were really to my satisfaction. Concepts behind the blueprints were creative enough and they all looked quite pleasant on paper. However, each and every finished building invariably disappointed, even perplexed me. In the year 2007, I was fortunate enough to study at TUM (Technische Universität München) which gave me an invaluable access to a close look at the working methods of many European architects, those from Germany and Switzerland in particular. Lots of them work in quite small towns or villages, normally in little workshops with only 3—5 employees. They dedicated their whole life to designing small local buildings in peace and quiet. In this way, they got to obtain a deeper understanding of the local architectural knowledge and tradition through

实践中，更愿意不提概念，无关文化，抛却情怀，不叹乡愁，只用心地解决问题，如此念而不执，却感觉收获更多。

3 当项目进入施工阶段，你去现场的频率如何？通常会遇到什么问题，又是如何解决的？

我一般每10天左右至少会去一次。现场的问题往往是以下几类：一是看是否按图施工；二是选定各种材料；三是施工中遇到困难，需要调整方案；四是在现场发现原来的设计不够完美，需要进行修改。就建造而言，现场才是真正的老师，它教会了我太多东西。在玉华村厕所的项目中，有一片清水红砖墙，可是建成后不久，墙面就呈现出白花花的碱渍，那时我才知晓了若选用弱碱性红砖，或者用白色水泥都可以减少泛碱的程度；在鄣吴小书画馆中，我期待用席纹竹胶板做模板、以便留下席纹清水混凝土顶棚效果的时候，工人师傅说竹胶板席纹本身太浅、效果将不明显，建议在模板表面再铺一层农村晒稻谷的竹席，效果果然立竿见影；在几周前玉华村委会的

frequent and effective communications with every aspect of the construction. Working locally gave them easy and frequent access to construction sites for more hands-on supervising. To them, high-quality designing and construction are common and natural. It was from then and there that I decided to make "remain local, design small" as my ultimate pursuit and goal. And then, it so happened that a string of luck and coincidences got me to start on my current projects at Zhangwu Town. At the very beginning, I took the job purely out of a whimsical interest, but as it progressed, those trivial and fragmented ideas gradually coalesced and integrated into a complete set of methodology and philosophy of my own. And that is what I call "Bring architecture back to the ground: my teaching and designs for life and land." I think it's safe to say that this particular piece of land at Zhangwu has, to a great extent, changed the way I thought about my teaching and working in urban architecture. It has even caused a fundamental change in my attitude toward life itself. Now, I tend to shun away from those lofty concepts, or meaningless ideologies, and just focus on solving the problems and getting the job done without any unnecessary overthinking. It gives me a greater sense of accomplishment.

参与村里的茶叶采摘 | Participating in Local Tea-leaves Picking

工地现场，建筑主体框架已经完成，还未砌筑墙体，站在门厅，我惊讶地发现透过内院能清晰地看到不远处极好的山景，于是深感有必要把原来的窄窄的带型窗改为大片落地窗……在现场除了解决施工中的问题，我惊讶地发现与业主、施工、监理等各方的关系也在悄然改变，大家在讨论、交流之中，变得越来越相互理解、包容，也越来越相互尊重。在一遍遍地往返现场之后，我也认识了越来越多附近的居民，以及各类店家的主人，以至于我现在每次去现场，都有回家的感觉，也深刻感受到只有把自己变成了当地人，才有可能真正在那里建一个好的房子。

4 “乡建”热潮背后不乏投机，你认为“乡建”能为乡村带来哪些实质性的改变？

诚然，“乡建”热潮背后中有投机，“投机”固然不好，但是“机会主义”我想至少应该是中性的，因为我们每个人或多或少都是“机会主义者”。在火热的“乡建”背景下，把握住各种机会，是机缘、毅力，也是能力。建筑师在乡村里，只要是面对真实的问题，提出了解决的方案，就值得点赞。很多人认为乡村设计房子比在城市更加自由、容易，其实未必，因为乡村建筑有着更加明确的投资、使用人群，有着更加注重自身利益边界的邻里，有着更有限的投资和薄弱的技术力量与施工手段。与城市中资本与权力往往过于强大的状况相比，乡村中各方力量之间相对更加均衡，也意味着更多的相互制约，所以在乡村建造中，你必须接受所有的限制，不能想做什么就做什么。

就整体而言，我不知道“乡建”究竟能为乡村带来哪些实质性的改变。当下对于乡村来说，最需要重塑的显然不是建筑，而是社会。我认为，乡村建筑，在当下作为一种机制与方法，而不是一种固有风貌或审美对象，其意义远胜过其空间形态和风貌本身。让建筑重回场地，让建筑重回生活，让建筑师重返建造现场，让建筑从形式本位中解放出来，或许这才是当下乡村建造之于建筑教育与建筑实践的意义。

3 How frequent do you visit the construction site? What are your most frequently encountered problems and how would you solve them?

Normally, I visit the construction site once every 10 days. All the possible problems I would have to deal with roughly fall into the following four categories. First, I have to make sure that everything goes according to the blueprint. Second, I need to check if the right building materials have been chosen. Third, I have to come up with adjustment and remedial measures to any practical difficulties that might happen during the construction. And finally, it might turn out that my original designs are just not good enough and that's when a fundamental revision to the blueprint is in need. For me, construction sites are the real teachers from whom I have learned a lot. From the project of the public toilet at Yuhua village, I learned that to cut down on the effect of efflorescence (a chemistry reaction where the migration of salt to the surface of a porous material, where it forms a coating) which happened to a section of red brick wall causing the appearance of a layer of white coating on the surface not long after the construction was completed, I could simply choose a particular type of red bricks that are weakly alkaline or replace red bricks with white cement. During building a mini art gallery at Zhangwu, I learned another lesson from an ordinary construction worker when he suggested that to create a special pattern of Xi Wen marking on the decorative concrete ceiling, using bamboo plywood with Xi Wen marking as the stencil plate was not enough since the marking would be too shallow to leave any imprint. To fix this problem, we could apply an extra layer of bamboo mat on top of the stencil plate, which is usually used by farmers to hold grains for sun-drying, to strengthen the effect. Thanks to his wisdom and life-long experience, it worked perfectly. Just a few weeks ago, I was working at the construction site of Yuhua village council. As I was standing in the hallway looking at the newly installed wall-less building frame, a rather surprising finding struck me—you can get a clear eyeful of the astonishingly beautiful mountain landscape through the inner courtyard. Suddenly it hit me—to retain this amazing effect, it's necessary to cast aside the original narrow and strip-shaped window design and make it vast floor-to-ceiling glass window. Apart from the benefit of being able to deal with the countless problems on-site efficiently, I was amazed by the fact that my relationship with the property-owner, construction workers or the

5 最近读的有趣的书是什么？简单阐述理由。

作为一个教师，读书是每天自然而然的事情。我在学校的每一天，也几乎是从读书开始的。坐在充满了阳光的书桌前，一杯清茶，随手拿起放在旁边的一本书翻阅起来，时间虽然不长，但总是非常享受那短暂的安宁。我现在所读的书，主要有两类：一类是专业类的经典文献，特别是建筑历史，另一类则是反映现实社会与生活的文学作品。对历史，特别是近现代和当代建筑历史书籍的阅读，让我在教学过程中，有了一个相对完整的时空坐标，来与同学们讨论一个作品、一件事情的价值与意义。最近一年左右读过的文学作品给我留下深刻印象的当属金宇澄的《繁花》，《繁花》的谋篇布局与表达方式让我想起了斯卡帕的建筑。下面摘录《繁花》中的一段文字："阳光照进来，雪芝身体一移，绛年玉貌，袄色变成宝蓝、深蓝，瞬息间披霞带彩，然后与窗外阳光一样，慢慢熄灭，暗淡。阿宝停步说，我不是有意的，因为上班。雪芝说，我晓得。阿宝说，我不进来了。雪芝说，进来吧。阿宝不响。雪芝说，不要紧的……"小说《繁花》中尽是这样以旁观者的不带个人情感的叙述方式，超然沉静，平淡如水，但是在模棱两可、漫不经心的对话之下，世俗生活中的占有、利用、交易、欲望等悄然显现，特别小说中那上千遍的"不响"，淋漓尽致地展现了片段与留白的魅力。我想《繁花》的作者金宇澄先生应该是不知道建筑师斯卡帕的，但是两人在各自作品的结构编排、叙事方式、细节技巧等诸多方面却有着共通之处，在那一个个让我们铭记的建筑细节与生活瞬间之中，我们真切感受到了另一种生活，看到了另一种建筑。或许，这就是阅读的价值与力量所在，它让我们超越个体的局限、时空的疆界，牵引着我们在学术与人生的道路上看得更深，走得更远。

6 最近一次旅行去了哪里？有什么收获？

我不知道究竟该如何来定义"旅行"。身在旅途，因工作出差或单纯度假，其过程和结果是完全不同的，因为后一种的旅行总是让人浮在那

building manager, are gradually changing for the better. In constant communication and discussions with each other, everybody were showing more respect to each other as well. What's more, I got acquainted with lots of local people who are ordinary farmers and shop-owners, so that every time I visited the construction area, I felt like going home. The whole experience taught me that only when you "transform" yourself into one of the local people will you be able to build a really good house there.

4 There are supposedly lots of profit-seekers riding the current trend of "rural development fever". However, are there, do you think, any substantive transformation caused to the rural areas by this "country fever"? If so, what are they?

Frankly, there is definitely a factor of profiteering behind this trend which could bring damages to the rural areas. However, I still believe that "opportunity-chasing" is not necessarily a derogatory word. In fact, every one of us is more or less an opportunist. Therefore under this background, being fortunate enough to seize the right opportunity also shows that you have capability and perseverance. Architects who are contented to work in rural villages offering practical solutions to real problems are quite admirable. Lots of people consider it much easier for architects to design houses in rural areas than in cities. But honestly, that's not the case. The truth is that you can't just do whatever you want. More often than not, your hands are bound to be tied by all sorts of limitation and inhibitions. Compared with the sufficient resources of capital and power in cities, financial investment is more limited in rural areas. Technical supports and construction equipment are comparatively weaker. Rural buildings sponsored by fewer investors are often targeted to more specific groups of people so you would have to take into consideration the distinctive boundary of interests between different neighborhood.

On the whole, it's quite difficult for me to say what kind of substantive and sustainable changes will be brought to the rural areas by this development fever. I guess it' s still too early for anyone to come to a final conclusion. From my personal perspective, what really needs to be developed is by no means mere architecture, but the whole social and cultural landscape. I think thatrural architecture as a working mechanism and method should be more than aesthetics styles. Its significances are way beyond spatial form and physical features. For me, perhaps, its true impact on our architecture education and practice lies in the fact that it' s an opportunity to bring architecture from its lofty pedestal back down to earth, to ordinary life, and it also bring architects back to construction sites and release architecture from the old formalistic shackles.

5 Are there any interesting books that you've been reading and that you would like to share with us? Why do you find them "interesting"?

As a teacher, reading is a day-to-day routine to me. Every morning, I start my whole day of work by doing some reading. Sitting at my desk flooded with sunshine, with a cup of tea close at hand, I would pick up any book that's lying on the desk and start enjoying the brief but cherished moment of peace and quiet. The books I've been reading fall into two categories. One is the classics in the field of architecture. I am particularly interested in architectural history. The other category is literary works that reflect real life and modern society. Reading about the history, especially the modern and contemporary history of architecture provides me a complete set of space-time coordinates which during my daily teaching, enables me to hold more in-depth discussions with and among my students about the value and significance of a particular architecture work or an related event. One particular literary work that I read about one year ago was a novel called *Fan Hua (All Those Flowers)* authored by Jin Yucheng, has impressed me most. Its distinctive style of story-telling and plot-construction reminded me of the Italian architect Carlo Scarpa and his distinctive architectural style. Here is an excerpt from the novel:

Streams of sunshine flooded in. Xue Zhi moved a few steps around the room. As she stepped through the lights, her face gave out a soft glow of silver and pink. The color of her coat instantly shifted from royal blue to navy blue, and then merged with the glorious sunbeam from outside the window, all bright and radiant, before everything faded away and back to darkness. A'Bao halted his steps trying to apologize, "I didn't mean to... Besides, there is work"... "No worries, " Xue Zhi replied, "I understand." A'Bao said, "I shan't intrude on you more. Got to go." Xue Zhi urged, "Would you like to come inside for tea?" A'Bao remained silent, still hesitating. Xue Zhi assured him, " There's no trouble. It's all fine."

The whole novel is full of this objective narration revealing no feelings on the part of the characters or the author, which creates a quality of aloofness and tranquility, as calm as still water. And yet, beneath the careless conversations of ambiguity lie plenty of desire, possessing, manipulation, and exploitation happening constantly in real life which keep emerging and showing themselves for readers to see. What's more interesting is the repeated appearance of the words "remain silent" which fully shows the power of "segmentation" and "leaving a blank", two unique features of Chinese art. I don't think the author Mr. Jin Yucheng knows anything about Scarpa the architect, and yet, they have so much in common their respective works in terms of structuring, plot designing, detailing techniques and many other aspects. One after another memorable designing details and small moments in life have given us a deeper feeling about a different life. They enlightened us to a different type of architecture. This, perhaps, is the true power and value of reading. It enables us to go beyond our individual limitation, and across the vast expanse of time and space, leading us to a deeper and farther reflection on our life and academic pursuit.

6 Would you like to talk about your latest travel experience? Where did you go? Any new thoughts or accomplishment?

I'm not sure how to define "travel". Depending on whether you are on the road for work or for vacation, its process and results can be absolutely different. Traveling for pleasure can never allow you to get yourself exposed to the local people and their daily life. You simply float in and out like clouds without ever touching the ground while thinking the picturesque scenery and its booming city in front of your eyes are all there is to life. So that's why for every holiday I choose to stay at home with my family. Everything we do is within a one-kilometer radius from our house, whether it's shopping, walking around a snack street, hill-climbing or simply looking out of the window admiring the flourishing car-washing business in our neighborhood. These sorts of stuff are not exactly exhilarating but they sooth me and fill my heart with peace and calmness. If I must say, the latest long-distance travel I made was a work trip to a small village called Huo Shi Shui located within the city of Qing Yuan in Guangdong province. During my stay I met a 96-year-old man. He is the father of the village head. The first time I saw him he was just walking back home from the mountain. With quite a heavy load of firewood on his shoulder, he took every stride with ease

个地方之上，让你无法真正接触那片土地和生活在其间的人们，误以为眼前的美丽如画的风景与城市就是生活的全部。所以每逢假期，我和家人一般是不出远门的，大多只是在以家为中心一公里左右的半径范围活动，去集市、逛小食街、爬附近的小山，看着门口生意火爆的洗车场……那种感觉不是让你激动欣喜，却让人的内心安定自在。如果一定要说个最近去过的比较远的地方，那就是我因工作的原因刚去过的广东清远的乡村，在那边一个名叫活石水的村里，我见到了村主任96岁的父亲，那天他正扛着一捆柴火从山上归来，神情刚毅，步履平稳。村主任说他父亲上午还独自赶集归来，而且其生活的状态几乎每天如此。看到那一幕，我心里涌出许多感慨：生活诚然辛苦，可也成就了老人圆满的一生！人在旅途，其根本的意义不就是为了让我们超越自己的身体和心灵的局限，向前能够走得更长远一些吗？“步数”固然重要，“远方”着实令人向往，但如何走得从容、自在、长远，才是旅行中我们真正需要思考的问题。

7 是否有对你特别有启发的建筑师？请简述理由。

实在是太多了，早期所喜爱的建筑师往往是因为他们对于空间与形体的塑造，而近些年深深打动我的却是他们的生活方式与实践理念，有三个人特别值得一提。第一个是西班牙建筑师巴埃萨（Alberto Campo Baeza），在业界享有盛名的他将事务所蜗居于马德里市中心一个居民楼顶层狭小的空间里，没有任何标识，其团队在过去的几十年中始终只有三个人。我问他为什么不做得更大，因为以他如此高的声望，这将是一件极其容易的事情。可是巴埃萨说，他不想把时间花在人的管理之上，而更愿意专注于建筑本身。巴埃萨再一次印证：一个建筑师工作与生活的方式，在很大程度上就决定了建筑实践的品质与特征。所以如果我们建筑的品质出了问题，那很有可能是我们建筑师自身的工作与生活状态有问题。第二个人是我在慕尼黑工业大学学习时的导师拉茨（Peter Latz）教授。他常说的一句话是：“你不

and confidence which is quite exceptional for his age. His son told me he took a long walk every day, be it to the town market or the mountain. I was both impressed and inspired. Though filled with disappointment and hardships, his life turned out to be equally fulfilling and successful. In fact, the very meaning of the life journey is for us to break through the physical and mental boundaries constantly and be able to go further. How many steps and how far you can go are admittedly important. However, the question we all need to really think about is how we can take our stride with more ease, comfort and freedom.

7 Are there any architects who are particularly inspirational to you? Why?

There are so many architects who have inspired me one way or another. In my early days I admired them for their exceptional architectural styles. However, in recent years, I tend to be deeply impressed by their lifestyle and practicing philosophy. Three particular architects are worth mentioning here. The first one is Alberto Campo Baeza, a Spanish architect. Though enjoying quite a prestigious reputation in architecture field, he chose to keep his office hidden in a tiny and plain-looking attic on the top floor of a residence building at downtown Madrid, without even a sign on its door. Over his decades-long practicing, he has always kept a small team of three people. I once asked him why he never thought to expand his business, which would be quite easy given his great reputation. He told me that he didn't want to waste any time on personnel managing which would get in the way of his real work—architectural designing. To me, Baeza is another proof that an architect's work and lifestyle can, to a great extent, determine the quality and features of his or her work. Therefore, if problems occur to the quality of our architecture, it's probably because there's something wrong with the work and life style of our architects. The second architect that I admire is Professor Peter Latz who is also my advisor at TUM. One of his frequent remarks goes like this: "Don't rush to think about any details about shape or space. What you need to do first is try to establish an inner connection between items. Grammar is way more important than those fancy words. It's like nature. Once a set of fundamental operating mechanism has been established connecting four seasons as well as days and nights, nature will present myriads of ever-changing scenery." That's why I often caution my students

要去设计具体的空间和形式，你要做的事情是建立事物之间的联系；语法远比华丽的话语重要，好比自然界，一旦建立了春夏秋冬、昼夜交替这样一种基本的运行法则，便自然演绎出千变万化的风景。”所以，我现在常常告诫自己和我的学生，别把建筑看得那么重要，因为那只是一个开端。在时间的沉淀、生态的蔓延、人的活动填充之后，一个建筑才可能真正成熟与完美。第三个人则是台湾建筑师黄声远。打动我的与其说是他的作品，不如说是他那种具有某种理想主义的生活与实践方式。也正是这种方式带给他许多不一样的作品，给这个越来越模式化、商业化的建筑世界注入了些新鲜气息。与黄声远的交谈也着实令人愉悦，就像是碰到一位多年不曾联系的老友，能感受到他那份源自心底的坦诚、洒脱与自在。在他身上，我看到建筑已是一种人生的修行，这种工作的状态与境界让我崇敬。

8 在常规学院教育中，你最喜欢的是哪一门课？目前，建筑教学中，你认为哪些部分是最迫切需要改进的？

我本科和硕士的学习都是于西安建筑科技大学完成的。当初选择建筑学专业纯属偶然，专业课的成绩也都只是平平而已，谈不上对哪门课最喜欢。当初特别羡慕那些能画一手好的钢笔、水彩画的同学，以为一张精美的手绘效果图就几乎是建筑设计的全部。客观来讲，本科阶段的学习于我个人而言只是建筑入门而已，所以我现在也常常提醒自己，对于那些“建筑感”不好的同学要给他们时间，也常常鼓励那些美术不好的同学——画画的能力与建筑设计的水平没有必然的关系。相比20多年以前，今天的建筑学和建筑教育早已经发生了巨大的变化。当下建筑学教育，在很大程度上脱离了土地与人，在一个相对封闭的系统里自我娱乐、孤芳自赏，或许这是其自身面临的最大困境，也是最迫切需要改进的地方。在即将到来的“无房可造”的年代，我们需要重新定义“空间”“形式”这些基本的概念，重新审视曾经头顶光环的建筑师这一“神圣”的职业。

and myself against stressing too much on designing and construction of a building because it's merely a beginning. A building can only become perfect and mature after the test of time, of human activities and of its ecological environment. The third architect is Mr. Huang Shengyuan who lives in Taiwan, China. What I find particularly admirable about him is not so much his works as his idealistic style of work and life. It is because of this unique style that enables him to create so many ingenious works which bring a blow of fresh air to this increasingly stereotyped and commercialized architecture world. Every conversation that I had with Mr. Huang was pleasant. It was like a chance encounter with a long-lost friend. I could actually feel his sincerity and ease emanating from the bottom of his heart. In him I can see that architecture has become "a practice of life", which I especially admire.

8 Within the regular course curriculum of your college, which course do you like best? And also, in the current architecture teaching content, which parts do you think are in desperate need of adjustment and improvement?

My alma mater is Xi'an University of Architecture and Technology. I got both my Bachelor's and Master's degrees there. I remember it was purely out of coincidence that I ended up choosing the architecture major. Back then, I wasn't exactly an A student, let alone any favorite subject. I didn't even have a clear idea as to what architectural design was. I thought being able to draw sheets after sheets of precision sketches is all it takes to be a qualified architect. I remember I was quite envious of my classmates who were particularly good at ink-drawing and water-color painting. Objectively speaking, my undergraduate study was no more than foundation laying. I only just stepped over the threshold of architecture. That's why I often remind myself of the necessity to allow more time for the students who haven't got a "sense" of what it means to be an architect. I repeatedly reassured my students for whom drawing is not a strong suit by telling them that there's not necessarily a link between your capability of drawing and architectural designing. Compared with the situation 20 years ago, the academic subject of architecture and architecture teaching today have been through drastic changes, not necessarily for better though. In fact, current architecture education has drifted away from the ground or the real life, and has been content with staying at an almost isolated corner

弱化建筑的物质实体功能，强化其介入社会生活的角色与方式，或许是让建筑学重生、建筑师重获地位的可能途径之一。

9 工作中有哪些印象深刻的教训？之后你做出了哪些改变？

工作中谈不上什么深刻的教训，但各种困惑与烦恼却还是有不少的，特别是源于自己的教师身份，在教学、科研、社会服务三大任务之下，受制于学科特点、学校考核、职业追求等，我曾处于一种极为困顿的状态，因为三条线很少如同所期待的那般彼此交叉融合，而多是平行，甚至矛盾的状态，这直接导致了作为教师身份的模糊与焦虑。在欧洲的学习与交流让我改变了对于职业与生活的态度，特别是这些年来，我有意拒绝商业性项目，聚焦乡村可以落地的小房子，探索建筑与土地、生活的内在关联。在一遍一遍乡村之行的过程中，曾经的某些小小抱负虽然渐行渐远，可是内心的方向却逐渐清晰、坚定，心态也相对自在起来。基于土地与生活的“在地建造”，于我而言，远未达到黄声远那种“人生的修行”高度，但不知不觉中它也开始成为一种生活的状态或追求。

for self-admiring. And that, I believe, is perhaps the biggest problem that's in desperate need of improvement. Facing the upcoming era when there would be not much space left for house-building, we need to redefine some basic concepts such as "space" and "form" and we have to switch to an different perspective to see the profession of architect that used to wear a "holy crown". One possible way for architecture and architects to regain their distinguished prestige is perhaps to dial down the physical functions of architecture and accentuate its role in social life and the way it plays this role.

9 Are there any unforgettable lessons you have learned in work? What changes have you made because of them?

"Unforgettable lessons"? Not much actually. However, there were certainly a lot of vexation and confusion. To be more specific, I used to be stuck in a state of weariness and perplexity resulting from my multiple duties, namely teaching, researching and social service as a teacher. Since these three lines of work are relatively parallel with each other, sometimes even in conflicts, I found it quite difficult to reconcile one with another in a harmonious mechanism and that directly caused my anxieties and puzzlement over the blurry boundary of my duties. Fortunately, my study experience in Europe at a later time completely changed my attitude toward work and life. In recent years, I have started to say no to those commercial projects and instead began to really focus on projects of small houses that are comparatively realizable and controllable in rural areas. Over my years of exploration in the inner connection between architecture, land and life, and during my repeated journeys to those small villages, I have been forgoing those grand ambitions or fancy aspirations. Instead, my inner compass starts to show a clearer and unwavering direction and finally I feel freer and more comfortable. For me, the practice of "adaptive construction" as an architecture mythology, which means "building based on the needs of sites and life" still has a far distance from what the architect Mr. Huang Shengyuan called "the practice of life", but it has gradually become a part of my current lifestyle as well as my life-long pursuit.

10 最近哪件社会议题最让你关注？你认为建筑的社会职能对设计会有哪些影响？

老实说，我是一个不太关注社会热点议题的，因为那些议题总是热闹地开场，悄然地结束。很多议题之所以成为“议题”，也不是因为它重要，只是基于一些机缘冒了出来，有很多根本就是“伪议题”。建筑的社会职能或属性对于设计当然有很大的影响，“传统与现代”“中国与西方”“建筑与文化”等等这类带有明显社会属性的主题探讨与相关思想深刻影响了我们在不同时期的建筑设计，但建筑品质的提高，最终依赖于建筑自身，那就是对于场地、材料、结构、建造、使用人群等这些基本要素的理解与把握。于我而言，相比积极关注社会议题，更愿安安静静地做自己可以把控的事情。

10 Are there any current social issues that you are particularly concerned with? What possible impacts do you think the social function of architecture will have on architectural design?

Honestly, I don't pay much attention to any social issues or "hot" topics since I believe they would always fade away pretty soon no matter how loud their starts are. The reason why many topics became "hot" at first is not because they are important or meaningful. More often than not, they became "hot" simply out of some kind of coincidence. In fact, many of the so-called "issues" are just "pseudo-issues". It's absolutely undeniable that the social function or social attributes of architecture has a great deal of impact on architectural design. The discussions and thoughts on topics like "Traditional versus Modern", "China versus the Western World" as well as "Architecture and Culture", which are obviously social, have deeply influenced the architecture styles at different period of time. However, we must also be aware that at the end of the day, the improvement of architecture quality still relies on architecture itself, which means the understanding and mastering of such basic elements as construction sites, building materials, structuring and building as well as the right group of

村里的早餐店 | Breakfast Shop in the Village

11 除了专业外，你还有哪些兴趣？你的兴趣是否与专业有所结合？

在学习、工作之外，我对其他事情一直没有显示出特别的兴趣，也没什么才艺或特长，总体上应该说是一个很“无趣”的人。我现在的大部分时间其实都是在各种工作中度过的，今年春节以后，女儿记录的每天家庭做饭的表格上，我只做了两顿饭……我不知道这样到底好不好，但我知道现在的工作，无论是教书还是设计，的确给了我更多信心和勇气。

12 送句话给将要毕业的同学吧……

每次同学毕业的时候，我都会扪心自问，我们真的对得起这批最优秀的年轻人在这里度过的生命中这段最美好的时光吗？就建筑设计专业技能而言，大学现在所能教授的其实越来越少，因为知识传播的方式与渠道相比过去已经完全不同。或许现在及将来大学能给予同学们更多的是关于职业与生活的观念、方法与态度。走出校门，对于每位同学而言，生活才刚刚开始。世界很大，空间也很多，大家要尽早在其中找到自己的“位置”，让内心安定，做自己觉得真正有意义的事情。如果要用一句话概括，我想说：先忘掉设计，专注生活，唯有如此，才能真正设计建造出好的房子！

users. For me, what's really needed to be done is not watch out for those social issues constantly, but rather keep calm and carry on doing what you can really control.

11 Apart from architecture, your own specialty, are there any other fields that you are particularly interested in? Are they connected with architecture?

Actually, I hardly have any particular interests in anything outside my own work and study. I'm one of those talentless "boring" person. A majority part of my time has been dedicated to my work, which just come my way one after another nonstop, so much so that I don't even spend enough time with my family. For example, on my daughter's "kitchen duty" chart, the number of cooking I have done for my family since this Spring Festival is a meagre "two"··· I know this is not something to be proud of. However, one thing that I'm sure about is the fact that my current work, be it teaching or designing, indeed has been giving me more confidence and courage.

12 Any parting words for your graduating students ?

Upon every graduating season, I would ask myself the same question: Is the education we have offered here to this group of the brightest young students really worth the dedication of the very golden period time of their life? In respect of the professional skills in architectural design, what the university courses can actually teach the students is not that much really, simply less and less, because the methods and channels of knowledge transmitting have completely changed. Perhaps the current as well as the future teaching that universities can provide to their students would concern more with abstract ideologies and methodologies. To every student, life only just starts when stepping out of the school gate. What's waiting for them out there is a much bigger and wider world. So I would advise them to try to find the direction and place of their own as soon as possible and find out what they feel are truly meaningful to do and get on with it. Only then can they really settle down and find their inner peace. In short, I would say: "Only by forgetting about those lofty and seemingly fancy concepts in architectural design and simply focus on experiencing real life, can you create and build truly wonderful houses."

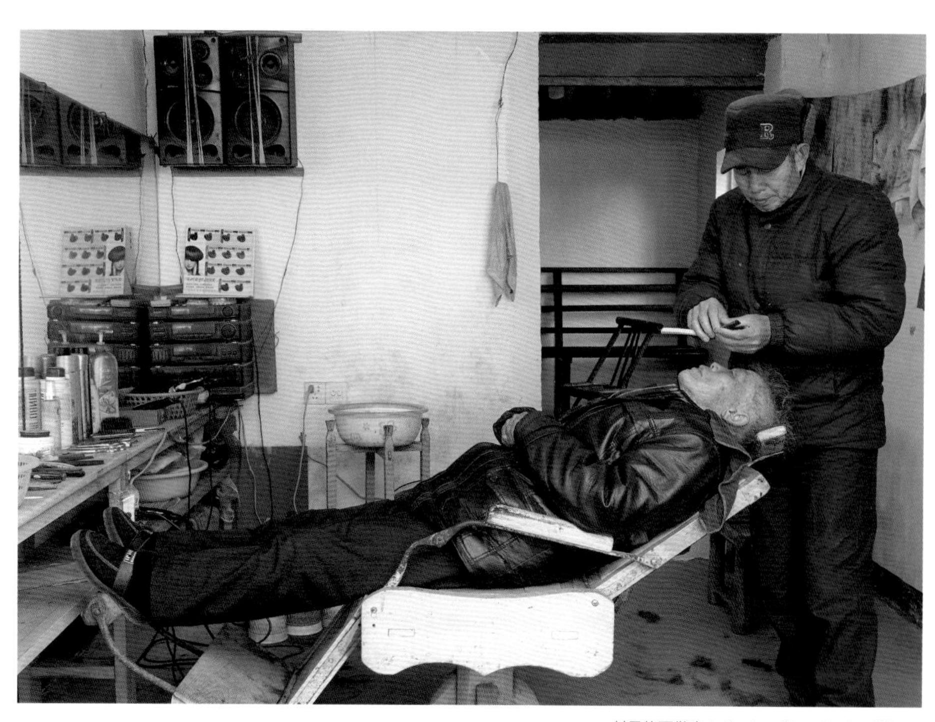

村里的理发店 I Barber Shop in the Village

观点 2 **来自鄣吴的消息**

Viewpoint 2 News from Somewhere Called Zhangwu

青锋，清华大学建筑学院 | Qing Feng, School of Architecture, Tsinghua University

浙江安吉县鄣吴镇鄣吴村的村头是一座小车站，仅仅由一个候车亭与一个分离的卫生间组成。建筑师贺勇在这里设置了两个巴拉甘风格的房间，分别粉刷成鲜艳的红色与蓝色，房间顶部垂下一道方形天窗，强烈的光线让色彩弥漫整个空间，浓重而纯粹。但最令人惊讶的是房间的功能，它们竟然是一男一女两个厕位，沉浸于巴拉甘式氛围中的水箱让人想起杜尚 (Marcel Duchamp) 的《泉》，只是这里的小便池是真的要作为小便池来使用。

要将巴拉甘著称于世的“宁静”与车站厕所的实用功能结合在一起，对于任何建筑观察者来说都不是一件容易的事情。在我们通常看来，巴拉甘花园中的沉思者与鄣吴村需要解决内急的旅客是完全两个世界的人，他们之间唯一的联系是贺勇，一位乡村建筑师和大学教授。这两个房间更像是他对经典的致敬，而非来自村民的日常习惯。

贺勇的做法显然不同于今天常见的乡建模式。后者往往侧重于乡土建筑类型、材料、建构特征、手工艺传统的尊重与挖掘。如此“简单粗暴”地将一种异类的“精英”建筑语汇强加在乡村生活之上，可以被轻易地被指责为对“文脉”的忽视，而排除在乡建主流之外，更极端一点甚

Zhangwu Village is located in Anji County of Zhejiang Province. At the entrance of the village lies a small bus station which consists of a waiting shelter and a separate public toilet. The public toilet is divided into two rooms, both of which are in Luis Barragan style painted in two different bright colors respectively. One is red and the other is blue. There is a square skylight framed in the slope of ceiling letting in sufficient daylights which would flood the whole room with colors, dense and pure. What really amazes me is the setting of the toileting area. There are two stalls, one for woman and the other for man which is quite unusual. The water tank is immersed in a Barragan ambiance which, in a way, reminds me of the urinal in one of Marcel Duchamp's masterpiece, Fountain. Except the urinal here is not just an art piece for people to admire from a distance.

To integrate the practical function of a public toilet with Luis Barragan's famous concept of "tranquility" is not an easy thing to do for any architecture observer. For us the meditator in Barragan's garden and the ordinary passengers at Zhangwu Village who are in urgent need to use the toilet are from two totally different worlds. And yet, here they are somehow connected by He Yong, a country architect and university teacher. These two rooms are more than just an answer to the villagers' daily needs, rather a homage to the classics of architecture.

He Yong's work is clearly at odds with the current popular model in rural development. The latter leans towards inheriting and developing

至可以被贴上“负面案例”的标签。但另一方面，“反潮流”的特征也恰恰提醒我们差异性路径的可能性，这需要对鄣吴镇传递来的消息做更审慎的了解与判断，再去讨论它的价值或局限。

接受与改变

巴拉甘式厕所展现了贺勇的这些乡建作品中的一种张力：两种氛围、两种传统，乃至于两种世界之间的吸引与排斥。不能简单地用“融合”来掩盖冲突与矛盾，需要观察的是这种张力所能带来的运动与变化。特定的物理学定义也适用于对建成环境分析。这种观察会将我们引向鄣吴村这几个项目中最有趣的一些地方。

单独地看，我们很容易怀疑建筑师在一个厕位上兴师动众是否过于小题大做，甚至会对建筑师过于强烈的个人印记感到忧虑。但如果对贺勇的其他项目有总体的了解，就会理解为何最强烈的建筑手段会出现在最不起眼的角落。这实际上是一个总体趋势的极端体现。在贺勇这些乡建作品中，项目越是重要、公共性越强、价值越高，建筑师的控制力就越会受到限制；反之亦然，建筑师发言权在边缘地带更容易受到尊重。我们有足够的例子来予以证明。

在厕所外的候车亭，是更为公共的场所。建筑师把一长排毛竹竿挂了起来作为隔断使用，有风的时候毛竹互相碰撞会发出阵阵声响。竹子是安吉特产，但投入使用后不久就发现并不稀有的毛竹竿却日渐稀少，原来是一些下车的乡民会顺手扯下竹竿当扁担把行李挑回家去。显然建筑师并没有预料到这种情况，他更没有料到的是这个新建的候车亭并不能作为正式的站台使用，因为旁边的土地问题，无法拓出足够的回车场。贺勇只能在一旁另行设计一个站台，我们去看的时候，候车亭二期正在施工。

这些意料之外的事让小车站的故事变得饶有趣味，村民难以预测的反应极大削弱了建筑师的“独断”色彩。建筑师有自己的意图，希望村民接受，而村民也用他们的方式去对它进行改变，这并不是一种对抗，更像是建筑师与村民之间一

the traditional rural buildings with its own unique structure and materials as well as the craftsmanship whereas Mr. He forcefully imposes on the rural life an alien "elite" architecture vocabulary in such a seemingly simple and crude way. His work may easily be accused of abandoning our "culture heritage" and excluded from the mainstream or worse, be labelled as a negative example. However, from a different perspective, this kind of "counter-trend" feature is also a reminder for us of a possibility of alternative methods which are in need of a deeper understanding and more discrete judgement to the messages from Zhangwu before discussing about its value or limitation.

Receiving and Altering

The Barragan style toilet is a showcase of the unique tension in He Yong's work between two different traditions creating two different atmospheres, which, furthermore, reflects the attraction as well as rejection between two different worlds. It would be unwise to simply cover up this conflict with "integration". Instead, we need a close observation of any potential motions and changes brought by this tension. Here we can borrow certain concepts from physics to do some analyses for the finished building and its environment. This kind of observation will lead us to the most interesting pieces among the projects at Zhangwu.

Seeing it separately, we would easily come to the conclusion that the designer is making mountain out of a molehill in a small toilet. Some of us even worry that the design bears too much of the designer's personal signature. However, if you have seen his other projects, you would understand the odd existence of so drastic a design approach at such a corner of obscurity. As a matter of fact, this is a extreme embodiment of a general tendency. The more public, expensive and important a project is, the less control the architect could have over it and vice versa. Therefore, architect's personal traits will be much easier to be respected and preserved, for which we can find plenty of proofs.

Now let's turn to the bus waiting shelter which is a more "public" facility. The designer has a long row of bamboo poles hung up functioning as a blocking screen. Whenever there's wind in the air, those bamboo poles would bump into each other creating a melodious sound. Bamboo is a specialty of Anji County and therefore should not be in short supply. Strange enough however, not long after being installed, the bamboo poles started disappearing

个的游戏。

另外两个项目更为典型地体现了村庄业主的干预。一个是景坞村旅游接待中心，位于村口小广场上。贺勇的设计是一系列单坡顶白色小房子，以不规则的布局散落在广场边缘。因为位置方向的差异，小房子之间会出现不同尺度与形态的户外空间，以此可以模拟村落场所的灵活与丰富。除此之外的一个主要元素是一条环绕整个场地的混凝土顶连廊。在南方的多雨气候中，它为室外停留提供了很好的庇护，也在错落的白房子上留下多样化的光影效果。对于一个乡村社区中心来说，这个设计的尺度、氛围、造价都是适当的。同样，不可预知的事情发生了，因为是村里的重点工程，甲方的意见变得格外强硬。最后完成的状态不仅舍弃了混凝土顶连廊，还给每个小房子粉刷了饰带，这是乡村建筑外部装饰的典型“官方”做法，但是贺勇最初设计中的质朴、纯粹以及虚实对比也都荡然无存。

另一个项目，鄣吴村书画馆也同样具有特殊的重要性。鄣吴村擅长制扇，又是吴昌硕的故居所在地，近年来的文化旅游开发投入不小。好在村子里还保留了传统的巷弄、水道肌理，一些民居也仍然是青瓦白墙的老样子，江南村庄氛围在某些地方还很浓郁。新建的书画馆位于村里的核心地段。贺勇的设计与景坞村社区中心的策略类似，两座白色小楼成 L 形布局，平面形态、相互关系、开窗位置与比例都旨在延续旁边传统民居的生活逻辑。两栋小楼之间是一座小茶室，一道楼梯环绕茶室上行，可以从二层进入书画馆。混凝土连廊再次出现，它围合出一个小院，一棵大树给院子足够的阴凉。从图纸看来，设计接近于阿尔瓦罗 · 西扎（Alvaro Siza）早期的设计策略，尊重历史场地的传统限制，挖掘日常的特异性，纯粹的白色墙面作为克制的背景使上述元素更为鲜明。书画馆灵活的流线、虚实边界的变化给予这个小建筑充分的内容与细部。在原来的规划中，书画馆的对面还有另一个新建的二层文化设施，与书画馆一道围合出一个小广场，成为村里为数不多的公共空间。这样一个核心文化设施，

one by one. As it turned out, some villagers who just got off the bus and were in need of a shoulder pole to carry their luggage back home would take one of the bamboo poles without realizing its true purpose there. Obviously, it's an unexpected outcome. What's more surprising for him was that due to the scarcity of space, this newly-built waiting shelter could not be used properly because there's no turnaround space for buses. Eventually, Mr. He had to design and build another bus station at an adjacent location. Last time we visited, the construction was just underway.

It was the unpredictable reaction from villagers that have made it impossible for the architect to completely "have it his way" and that is what makes the story of this tiny bus station intriguing. It's more like a game than a conflict between the architect's intention and the villagers' unwitting modification to it.

We can find a more typical example of intervention from rural land owners in He's two other projects, one of which is the tourist center of Jingwu village. The tourist center is located at the public square at the entrance of the village. The original plan was a series of white cottages with single pitch roof, scattering randomly around the edge of the square. The variety of distances between the houses and their respective frontage are supposed to create outdoor spaces of different size and form for each little house, showcasing both the flexibility and abundance of the rural land. Another main element to the design is a concrete-roofed corridor winding around the whole square. Considering the typical rainy weather of the southern China, the roof is quite a fitting idea as it provides shelter for a temporary stay. In addition, the whole setting creates an ever-changing interplay between lights and shadow. For a tourist center of a small village, it should be an appropriate plan in terms of its size, cost and surroundings. And yet, once again, something unexpected happened. Since it was one of the "key projects" for the village, the owner's suggestion and requests were especially needed to be accommodated. Eventually, the architect had to forgo the corridor altogether and add the painted ornamental ribbon band, typical of rural buildings on the facade of each cottage. Regrettably, however, this modification totally cancelled out the simplicity, purity as well as the contrast between void and solid presented in He Yong's original design.

The other project was the art gallery at Zhangwu village. As part of the tremendous amount of investment in its culture tourism development

自然更受“重视”，最终只有书画馆得以建成，总体格局仍然遵循原有设计，但是细部的调整，如青石板墙裙、披檐门斗等“典型”做法的加入剥夺了原设计中微妙的不寻常之处，而这本是设计策略中所依赖的催化剂。与景坞村类似，原设计的“正常”化修正剥离了不少建筑师精心考虑的细节，这种结局同样来自两个世界的碰撞，一个奉行“上帝在细部之中”的毫厘雕琢，另一个是“官方常规”的名正言顺。

现在回看贺勇的巴拉甘式厕所，最初的疑虑甚至可以转化为某种程度上的同情，只有在这最为私密的角落，建筑师的意图才能得到最大的保全。而越外向、地位越高的地方，主导权也越多地受到村庄业主的节制。考虑到社区中心与书画馆中，原有设计品质因为改动所遭受的影响，小厕所的“小题大做”变得多少可以接受。不管将它视为遗存还是补偿，在总体图景之下，它从另一个侧面体现了鄣吴村乡建中建筑师与村庄业主的关系。建筑师不再独自站立在舞台中心，另一位主角——乡村的身影甚至更为强大。

垃圾站与小卖店

接受村庄成为主角的价值之一，是主角们都会有兴趣把对手戏继续下去，获得机会的建筑师也有可能将剧情带向不同的方向。贺勇的另外两个项目，鄣吴镇垃圾处理站和无蚊村小卖店就是这样的剧情转折。

相比车站、旅游接待中心与书画馆，垃圾处理站与小卖部的关注度要低很多。按照此前总结的规律，村庄业主的干预会小很多，建筑师的自主性相应增加。实际情况也的确是这样，贺勇的设计基本能够较为完整地实施下去。而在建筑师这一面，经典建筑语汇仍在出现，但也不同于车站厕所中那样强烈反差。这两个项目中展现了主角之间不同的相处方式，以及随之而来的不同结果。

鄣吴镇垃圾处理站位于村外的小山坡上。原有垃圾房仅仅起到临时堆放的作用，此后垃圾站增加了分拣处理功能：厨余垃圾进入发酵处理器被转化为农用肥料，剩余的生活垃圾经过机器压

in recent years, this project was also especially important. One of the cultural specialties of Zhangwu village is its unique crafts of fan-making. Zhangwu village is also the hometown of Wu Changshuo, an prestigious Chinese artist during the late Qing dynasty. Those are the reasons why the whole village is shrouded in a traditional artistic atmosphere typical of Chinese Jiang Nan (literally the south side of Yangtze River) villages. Here, you can see the narrow alleys, the petite waterways as well as the old-fashioned houses with white-painted walls and green-tiled roof. The new art gallery is located right at the center of the village. His approach to its design was similar to that of the tourist center mentioned above. There are two small white buildings standing next to each other forming "L" shape, connected by a small tea room in-between. Ascending the winding stairs from the tea room to the second floor, you will be standing inside the art gallery. The whole building complex has been designed in consistent with the very style of the residence buildings in the neighborhood from its lay-out to the positions and ratios of its windows. Once again, a concrete-roofed corridor has been used to surround the two buildings creating a small courtyard, at the center of which stands a big tree casting a cool shade blocking the all-too-bright sunlight. A closer look at the blueprint shows that the whole design looks similar to the early style of Alvaro Siza, a multiple award-winning Portuguese architect, in that the restrictions in preserving historical sites have been obliged and the idiosyncrasy of its local daily life has been shown. These elements are even more prominent against the modest background of pure white walls. Meanwhile, the streamline shape of the building edges as well as the fluid boundary dividing brightness and darkness offers plenteous content and details. In the original plan, there was supposed to be another culture facility, a two-storey building just opposite the art gallery, creating a mini-square serving as one of very few public spaces in the village. Such a core cultural facility naturally received more attention. However, for whatever reasons, only the art gallery got okayed for construction. Other modifications involved introducing the more common and popular blue flagstone dado and a door socket with eaves on top, replacing the architect's original design which is more delicate and unusual but functioning as some sort of "catalyst" for the whole designing strategy. Again, this outcome came from the colliding between two

缩装入垃圾箱中运走。垃圾站因此扩建了一座二层小楼。新旧建筑的布局完全由垃圾处理过程所决定，新建筑挖入山坡之中，二层地面与原有垃圾房齐平，便于通过传送带将生活垃圾传输到埋置于垃圾房地面下的压缩机中。一楼则放置厨余垃圾处理器，这样食物残渣可以从二楼地面的孔洞直接倒入处理器。

上下两层的不同功能直接导向了不同的建筑处理方式。新建筑上层体量更大，垃圾分拣时的臭味需要及时疏散，因此建筑师采用了空心砖、屋顶开缝等元素，并且将混凝土结构与灰砖砌块墙体直接暴露在外。意料之中的是，村里没有再要求给予白色粉刷和灰色饰带的优待。下层完全是另外一种氛围，房间内铺有地砖，墙面白色粉刷，一台整洁的不锈钢处理器占据了半间屋的面积，竖条窗和木板门都在提示这是一个房间，不同于楼上的车间。上下层功能与气质上的差异性也体现在室外。与上层粗糙和直白的灰色形成对比的是，建筑师在下层采用了红色黏土砖与面砖来铺砌地面、墙面、坡道与台阶。江南的雨水很快就在砖砌踏步的砌缝间培育出翠绿的青苔，让建筑师的意图一目了然——以红砖的温暖和拙朴营造一个亲切和平静的角落，旁边的竹林与门前的池塘也是这个景观设计的一部分。

不难理解这种差异所传达的讯息。上层的分拣、传送、压缩属于机械流程，建筑师相应地把结构与材料最直接的样貌暴露出来，效率与合理性是核心的诉求。下层的处理器实现一种特殊的转化，厨余垃圾被转化成肥料，最终又回到村里的土地中去。确实没有什么材料比红砖更有利于陈述这种有机循环的理念，我们不难在芬兰建筑师阿尔瓦・阿尔托 (Alvar Aalto) 的作品中找到相似的场景。象征性诠释的延伸是这所小建筑鼓励人们去感受和解读的。

这个小建筑之所以值得专门讨论，在于它的“功能”实现超乎想象。我们去参观时，刚刚完成了垃圾的压缩，整个垃圾站竟然看不到一点裸露的垃圾，地面、墙面与机械也都保持洁净，这与我们平常对乡村卫生条件的不满形成了强烈的

different worlds, one being the precision carving in the belief of "God is in the details" and the other following the official regularity.

With that in mind, now it would be easier for us to understand the very existence of Barragan style toilet—only at such a modest and private corner can the architect's original intention be preserved to a great extent. Meanwhile, the more public and conspicuous a project is, the more control and intervention it will get from the landowner, for which we can't help but feel a little sympathy for the architect. For in this sense, the architect is no longer at the center of the stage. Instead, he has been overshadowed by another leading actor, the village authority. Considering the amount of changes imposed on the tourist center of Jingwu village and the art gallery at Zhangwu village, we will be able to more or less understand why He Yong made such a big fuss over a tiny corner as obscure and humble as a public toilet.

Garbage Disposal Station and Convenience Store

Now that there is the second leading actor performing opposite him, the architect has been given the opportunity to lead the plot to a different direction. For He Yong, the plot twist turned out to be two other projects: the garbage disposal station at Zhangwu village and a convenience store at Wuwen village.

Compared with the bus station, the tourist center and the art gallery, the garbage disposal station and a small convenience store would receive less attention from the authorities, which, according to the pattern previously mentioned, means less intervention from the village landowner hence the relative increase in autonomy for the architect. And that is exactly what's actually happened for He Yong, for his original design has basically been preserved and carried out. This time, despite the recurring classical architectural vocabulary, the resulting style was not as drastic as that of the public toilet at Zhangwu bus station. This difference in outcome has resulted from the difference in ways that the two main actors interact with each other.

The garbage disposal station is located on a little hillside outside the village. The original station was simply a temporary landfill for the garbage to pile on. The newly-built part can serve as a recycling plant where kitchen waste will be put in a fermentation processor and turned into agricultural fertilizer while the remaining household garbage will be compressed by machine and packaged before being transported elsewhere. The whole process will be conducted in a two-storey building. The layout concerning the exact locations and the distance between the old and the new building has been decided based on the garbage-processing procedure. Therefore, the first floor of the new building was actually below the surface of the hillside leaving the second floor on the same level with the old house. This arrangement has made it easy for the waste to be conveyed from the second floor of the new building to the compressor which is installed below the surface of the old house. Meanwhile, the kitchen waste left on the second floor will be poured directly through the hole opened in the floor into the processor especially for kitchen garbage which is installed in the first floor of the new building.

The two different functions between the first and second floor have resulted in two different designing approaches. Since the second floor of the new building will be used for the garbage sorting center, the designer has made the room much bigger and he especially adopted the elements of air bricks and cleft ceiling for the purpose of fast dispersing the strong garbage odor as well as the concrete block and sand-lime bricks for the wall. Unsurprisingly, the village authority didn't ask for the addition of whitewash and painted grey decorative band as a special touch. The first floor, on the other hand, was a whole different situation. The floor has been paved with bricks, and walls are whitewashed. A neat processing machine made of stainless steel takes up half the room. There are also windows and wooden door. All of those are telling us that this is a room, unlike the room upstairs which is clearly meant to be a processing plant. The difference between the two floors in terms of function and style can be shown by the exterior of the building. In contrast with the ruggedness and plainness of the grey colored bricks on the second floor, the clay bricks and face-bricks applied on the surfaces of ground, wall, ramp and stairs for the first floor are all in bright red color. During the long rainy season typical for southern China, a layer of verdant moss soon grows out of the ever-moist interstices at the joints between bricks, which, coupled with the bright red color of the bricks, creates a simple and adorned space of peace and comfort. And that was exactly the architect's intention. Apart from that, the nearby bamboo grove and the pond facing the front door are also part of the whole landscape.

The message delivered by the contrasting style between up-and-down stairs is quite understandable. The sorting, conveying and compressing upstairs are all part of the whole mechanical process whose main concern is efficiency and rationality. So, the architect deliberately has the natural state of the building structure and materials completely exposed discarding all possible concealment or adornment. The downstairs, on the other hand, involves a specialized transformation from kitchen waste to fertilizer and finally back to the earth. This philosophy of organic circulation can best be expressed by none other than the red clay bricks. We can easily find the similar scenes in the works of Finnish Architect Alvar Aalto, which aims for the extension of a symbolistic interpretation for a building and that's also what this little building is trying to encourage people to understand.

The reason why this tiny two-storey building deserves a special discussion is that the effect and efficiency of its "functions" are beyond imagination. In our last visit, the whole compressing process has just finished. What surprised us most was that there's not a single piece of garbage littered around to be seen, leaving the surface of the floor, the walls and the machines all neat and spotless, in a drastic contrast with the usual impression we had about rural sanitary conditions which is "messy and stinky". Garbage sorting is another surprise to us. Even in a metropolitan city like Beijing, it's still at a stage of slogans and empty talk after so many years of advocating and yet at Zhang Wu, we have seen with our own eyes the collected household trash being transformed into packages of agricultural fertilizer. So you see, it only takes a small garbage disposal station to rectify certain prejudices against rural areas.

Apart from its practical efficiency, the interpretation to the meaning of "purpose" is also up to the architect. He Yong's approach to that often reminds us of all sorts of debates centering around the topic of "functionalism". In the early 20th century, the German architect Adolf Behne first made an distinction between "functionalist" and "utilitarian"[1], the former entailing a pursuit of meaning, and the latter only focusing on practicality and efficiency. Behne was undoubtedly a supporter of the former. In his opinion, by combining efficiency and meaning, a functional building can be linked with culture, philosophy, and even the shaping of human characters. Naturally, that would give the architect a bigger working space and richer resource of meanings. Meanwhile, in terms of

反差。垃圾分拣是另外一个意外，在北京这样的城市，垃圾分类宣传了很多年，仍然停留在口号与摆设。但是在郭吴，我们亲眼看到村里各家各户收集的厨余垃圾被转化为一袋一袋肥料。一个小小的垃圾站，足以修正对乡村管理与生活状态的某些偏见。

除去实用效能之外，对“功用”(purpose)的“意义”(meaning)诠释则要归功于建筑师。贺勇的处理很容易让人联想起围绕“功能主义”的种种争论。早在20世纪初期，阿道夫・贝恩(Adolf Behne)就将这种蕴含了意义诉求的“功能性”(functionalist)与单纯追求效用的“功利性”(utilitarian)区别开来[1]。贝恩所支持的当然是前者，通过效用与意义的结合与延伸，一个功能性的建筑可以与文化、哲学，甚至是对人的塑造相关联。这当然意味着建筑师更广阔的操作空间以及更丰厚的内涵来源。而对于后者，功能被缩减为量度的计算，枯燥与单一成为不可避免的宿命。遗憾的是，贝恩的精确分析并没有被大多数的人接受，“功能主义”几乎成为现代主义的原罪。

在贺勇的小房子中，上下两层可以被视为对“功能性”与“功利性”分别呈现。沿着这条思路，我们甚至可以将垃圾处理站与有机建筑传统，与表现主义建筑，甚至更早的浪漫主义思想联系起来。但这样显然会引发将“精英化”的理论体系强加于一个普通建筑的质疑，就像巴拉甘厕位的例子一样。但换一个角度看，为何要坚持“精英”与日常的割裂？这些体系之所以成为精英，恰恰是因为它们能够提供普遍性的、具有深度的解释，如果你不在自己的脑海中把它们当作“精英”而敬而远之，那么没有任何障碍将它们与一个垃圾站或者是厕所分隔开。精英与日常之间的差距或许不在理论与实践中，而是在人们自身划定的等级观念中。赫拉克利特 (Heraclitus) 在自己厨房中所说的话在今天仍然发人深省：“进来，进来！神也在这里。”[2]

需要我们软化精英与日常二元对立的情况，也出现在无蚊村月亮湾小卖店的设计中。这个小店原

the latter, "utilitarian", function is reduced to mere calculation and measurements, which unavoidably leads to monotony and simplicity. Regrettably, Behne's idea didn't get accepted by most people, so now, on some level, "utilitarian" has become an original sin of modernism.

In this little two-storey building designed by He Yong, differences between the upstairs and downstairs can be regarded as the separate representations for "functionalism" and "utilitarianism". In this sense, this garbage processing center can be linked to the traditional organic architecture, the expressionist style and even far back to early romanticism. Apparently, that would cause suspect and criticism to the act of imposing an "elite" theorization on an common and ordinary building just like the case of Barragan style toilet. However, let's look at it from a different angle. Why do we have to insist separating elitism from everyday life? Isn't the very reason why elitists are elitists because they can afford a general yet in-depth interpretation? The only real obstacle that stops you from associating the "elitist" theorization with an ordinary toilet or a garbage disposal center is your own alienation from "elitism". Perhaps the gap between the class of elite andcommon lies only in people's mind, which reminds me of what Heraclitus said in his kitchen: "Come in, come in, Gods are here too." [2]

Another case that is in need of softening this binary opposition between elitism and ordinariness appears in the design of a convenience store at the village of Wuwen. This was a renovation project since the old store was actually an unauthorized construction due to its inappropriate location, right beside the waterscape specially built by the village as a tourist attraction. The new building got to keep its spot but its design has to be in consistent with the waterscape in style. He Yong's design met the requirement and so got carried out, except one change—the skylight. It got removed because the contractor thought it was "not worth the trouble". Because of that, now the store will have to keep its lights on all the time due to the lack of daylight. Among all his rural projects, this little store enjoys the best condition in terms of location. It sits right in the converging juncture between three valleys where the streams flowing down from the mountain is stopped by a rock dam on three sides, transforming the once deserted rock patch into a rippling moon-shaped river bend. At the feet of the store, two streams meet each other and together they

来是村民搭的违建，因为处在村里特别打造的水景旁边，所以要进行改造，须兼顾小卖店的原有功能以及景观作用。贺勇的设计基本都得到了实现，唯一的改动是小卖店的天窗因为“麻烦”被包工头省掉，使得店里即使是白天也需要开灯补充照明。在贺勇的几个乡建作品中，小卖店的位置最为优越，这里是无蚊村中心三个山谷的交汇地带，从山里流下来的泉水被三道石堤拦住，原来的乱石滩由此变身为山光水色的月亮湾。小卖店就位于两条溪流的交角处，地势高出水面不少，两边都被水面环绕，经一道石梯可以从小卖店旁边下到水边，村里的妇人常常在此用山泉洗衣。

此处原有三栋小房子，错落布局，倒是很接近景坞村村委会的格局。这几栋小房子最大的不足在于面向月亮湾水面，过于封闭。过去这里是乱石滩时这并不是问题，但现在这里已经是月亮湾，做出改变也就理所应当了。贺勇的新小卖店仍然保留了三个房子原有的平面格局，由呈丁字

surround the store on both sides. At one side of the store, there are stone steps descending to the river and villagers often come over to do laundry.

In the place of the convenience store, there used to be three small houses with a similar lay-out with another project of He Yong's, the community center of Jingwu village. The biggest defect of these houses were their isolated positions due to the fact that all of them stand right next to the river bend facing the water. This was not a problem in the past when there was a rock patch, but now with the waters, changing is the only reasonable solution. The new convenience store has adopted the old lay-out with two rooms in perpendicular position to each other forming a "T" shape. The biggest modification the architect has made is transferring a formerly closed-up house facing the water to an open pavilion. The wall facing the road has been completely removed, creating a wide-open entrance for visitors. On two sides sit several Chinese traditional stools with backrest and an O-shaped window has been carved out of the wall facing the river bend echoing the elements of moon-shaped door or round window typical of Chinese traditional garden design. Learning from

村里制作竹扇的工匠 I Bamboo-fan Craftsman/woman in the Village

形相交的两个房间组成。坡顶元素也得到保留，在这里变成了平缓的单坡，从两端向中心汇聚。建筑师最大的改动在于将临水的封闭房子打开，转变成开放的亭阁。靠道路的一面完全开放供人进入，侧面设置了传统的美人靠座凳，面向河湾的墙体上挖出一个整圆的窗洞，呼应传统园林中的圆窗或者是月亮门。或许是吸取了村委会与书画馆的经验，这里没有再采取容易被“修正”的白色粉刷，而是用竹竿支模现浇混凝土来铸造这个小房子。竹竿与竹节在墙体上留下了很深的印记，爬山虎正在攀缘，一旁的翠竹与墙体上的凹槽形成巧妙的对话。

贺勇的处理显著地提升了小卖店的存在感。粗糙的墙面、统一的材质、连续的体量明白无误地呈现出房子的特殊性。圆窗则是最精妙的一笔，不仅给予了亭子鲜明的江南文化属性，它位于正方形墙体正中央的位置也渲染出一种含蓄的纪念性。无论是在东方还是西方，方和圆都被赋予和谐与永恒的寓意，贺勇再一次将一种经典理论传

the lessons and experiences of community center and art gallery, instead of whitewashing the walls which could be easily "rectified" by the authority, the architect made the walls by casting concrete cement into a special mould made by bamboo sticks. The visual effects of the finished walls are amazing. You can clearly see the markings of bamboo poles and joints on the surface of the wall creating a type of artistic grooves. Covered by the evergreen creepers climbing upward, the wall seems to be having a clever conversation with the verdant bamboo trees standing beside it.

He Yong's design has clearly put the store into a more prominent position. The crude texture of the wall surface, the uniformity of the material as well as the continuity of the form all represent the uniqueness of the house. The O-shaped window is an especially ingenious touch as it serves not only as the embodiment of a distinctive cultural attributes of southern China, but also an implicit metaphor, as in both Eastern and Western world, the shape of circle, represented by the O-shaped window itself, symbolizes harmony and the shape of square, represented by the square wall which the window is embedded in, symbolizes eternity.

村里的街巷 I Small Alley of the Village

统固化于小卖部的混凝土墙壁中。

这两种不同的内涵，可以解释我们面对这个简单的小房子时并不简单的体验。一方面是依山傍水的临泉小榭带来的惬意，另一方面是高居水面之上几何象征的纪念性。前者将我们引向山水之中隐现的亭阁，后者则让人联想起西方古典时代的神庙。将两种相对异质的传统并置在一起并不是第一次在贺勇的设计中出现，但在小卖店中应该是最为微妙的。如果从远处走近，最终进入亭子里面，就能清晰感受到两种内涵的转换。远观时水面开阔、地形凸显，“神庙”的纪念性更为强烈。走近一些，房子与周围草木石渠的关系展现开来，开始变得更为亲切。最后进入亭子内，透过圆窗看到对面的山水农宅，最终意识到你原来身处月亮湾，身处江南，身处自然园林之中。与垃圾处理站类似，这个小房子中也蕴含着两种话语体系的交融，只是在这里更为细腻与含蓄。

对于村庄来说，贺勇的小卖店起到了两种作用，补充了景观元素还在其次，更有价值的是，原来的小卖店仅仅是在地点上位于村子的中心，而现在的小卖店才是整个村子空间结构、场所氛围、活动交流的中心。亭子里的八仙桌与儿童游戏的摇摇乐透露出公共活动的频率。贺勇曾经希望景坞村村委会与鄣吴村书画馆能够成为乡村生活的中心，但我们去参观的时候，这两个建筑中几乎空无一人，反而是在无蚊村的小卖店，4位村民正兴致勃勃地玩着麻将，另外几位站在一旁围观。

乌托邦与乡村

上面讨论的5个项目并不是贺勇在鄣吴镇的全部作品，但是也足以传递一段值得关注的讯息：它们作为整体展现了一种特定的乡建模式，虽然不同于当下的乡建主流，但或许是更为真实，也更为现实的乡建模式。真实不仅在于这些项目已经建成，还在于它们是由乡村投资、乡村建造，并且为乡村所使用。现实则是指从设计到建造以及使用整个过程会受到很多因素的影响，贺勇几个设计不同的遭遇就体现了现实的复杂性。这两

The two different connotations offers an explanation to our somewhat complicated feelings toward this simple house. On the one hand, there's the coziness inspired by the proximity of mountain and river, which guide us to the half-visible pavilion snugly tugged in the picturesque backdrop. On the other hand, the commemoration of its geometric symbol standing above the water level conjures up the image of the holy temples in the Western classical time. This juxtaposition and combination of two distinctively different cultural traditions in one building has made its appearance in He Yong's works multiple times, but this particular one is the most subtle and delicate application. As you stroll leisurely toward the pavilion from a distance till arriving at its entrance, you could clearly feel the transition between the two different undertones. Looking from afar, you will see the vast surface of the river and the undulation of the landscape which conveys a feeling of solemnity and commemoration typical of a Western temple. Nearing the pavilion, everything has changed. The little house hugged by the surrounding grass, plants as well as the stone canal start to seem so ordinary and intimate. Finally, when you are actually standing in the pavilion, looking out of the round window and see the country house and its surrounding landscape, you start to realize that you are not in some stately temple, but in a little southern style landscape garden called "Moon Bay". Similar to the garbage disposal center, this little house also represents the integration of two different language systems. What's different is that the effect of the integration here is subtler and more delicate.

To the village, this little store serves two functions. One is supplementing the landscape. The other, and also the more valuable one is making the store the center of the whole village, not just in terms of the location, but also the center of spatial structure, atmosphere as well as social activities. The "Ba Xian" table, a Chinese traditional square table that can accommodate eight people, and the electronic rocking wagon for children clearly show that this is a place for public activities. He Yong's intention for the community center at Jingwu village and the art gallery at Zhangwu village was to be the center of the local public life. However, his plan didn't come to pass after all. Last time we visited these two places, they were almost empty. The situation at the convenience store at Wuwen village, on the other hand, was quite different. The little pavilion was actually fully occupied. There were four villagers sitting at the table happily

种特征都来自乡村在这些项目中所扮演的角色。在郭吴，乡村的立场更为强势，作为毋庸置疑的主角，其对项目的干预更为直接，它们更接近于我们通常所认知的“甲方”。

从这个角度来说，贺勇的设计所遭遇的其实是再正常不过的甲乙方拉锯，只是当这种拉锯发生在大学教授与乡村之间，发生在当下的乡建热潮中，反而变得有些“非典型”。近年来许多在大众媒体中传播很广的乡建案例中，乡村主要是作为一种背景出现，为建筑师提供不同于城市的场所环境、自然条件。乡村既有的建筑品质，如本地材料、传统建构、空间秩序往往成为建筑师最为珍视的设计出发点，由此才会有多种多样的特色鲜明的乡建成果。与郭吴的情况不同的是，这种“典型”乡建模式中，乡村被定格在沉默的“文脉”中，它以自己的传统为建筑师提供素材，而剩下的工作完全落入建筑师的手中。如何使用这些素材，用什么样的资源来完成建设，在大多数情况下，乡村的声音都是微弱的。因为主导了整个设计与建造过程，建筑师能够保证设计的品质贯彻始终，但潜藏的危险是项目虽然建造在乡村但是并不真正属于乡村，无法与日常的乡村生活相互融合。

当乡村成为布景，而不是建筑的切实发起者和使用者时，它也就会滑向阿道夫·卢斯(Adolf Loos)所描述的“波将金城”(Potemkin City)——一个刻意装扮起来的秀美村庄，其实只是为河对岸的女王观赏的假象[3]。卢斯以“波将金城”形容20世纪初的维也纳，那些被装扮成文艺复兴样式的建筑，仿佛能让建筑使用者瞬间变成贵族，实际上不过是自欺欺人。当下的一些乡建项目也在进行这种装扮，装扮的对象正是乡村，只不过是被“乌托邦化”的乡村。“乌托邦”往往是沉默的，因为一切已经完美，无法再予改动，也就不再需要不同的意见。在一些建筑师看来，乡村就是这样一个理想的“世外桃源”，它所拥有的自然条件、生活方式、建筑特色都是城市环境中所缺乏的，因此可以作为对城市生活缺陷的弥补，为那些对城市不满的人提供慰藉。这

playing mahjong with several onlookers standing behind watching closely.

Utopia and Village

The five projects discussed above are just a small portion of He Yong's works at Zhangwu, but they are enough to convey a message worth our attention. As a whole, these five projects represent a certain model for "rural development". It might be at odds with the mainstream, but it's perhaps the most practical and realistic one. It's practical not only because these projects have actually been built, but also because they have been invested, constructed and utilized by the local villagers. It's realistic because of the fact that these projects have been impacted by several factors during the whole process, starting from designing to construction and all the way to the day they were put to use, which reflected the complexity of the reality. Both of these two characteristics came from the role played by the local authority. For example, at Zhang Wu, the intervention from the local authority was more straightforward and unnegotiable. In a sense, they are playing the real leading role occupying the center of the stage. They are the real "Client".

From this point of view, the conflicts between He Yong and the local authority are nothing but the ordinary negotiations between Party A and Party B. They look "atypical" simply because these back-and-forth struggles happened between a rural village and a university professor. In recent years, a great number of "rural development" cases appearing in mass media features rural villages simply as a background, which provide architects with the working sites and natural conditions quite different from urban areas. The distinctive architecture traits, such as local materials, traditional building styles and the spatial order are especially respected and treasured by architects who treat them as the basic foundation for their design. That's why their works are so exceptionally diversifying and boast such distinctive features. Unlike Zhangwu projects, the mainstream rural building model put the rural areas in the traditional Chinese cultural context, serving as a silent material resource for the architect. Most of the time, the voice of the rural areas is feeble at the best. Architects are the ones who dominate the whole designing and construction process and therefore are able to guarantee the consistence of the quality of their designs. However, there's a hidden downside—the buildings can not really merge into ordinary country life. They do not really belong

样的乡建作品，建立在对乡村社会的选择性描绘之上，会过滤出有利于填补城市缺陷的元素，而其他的东西则与乡村背景一道沉入寂静之中。就像“波将金城”是为女王所准备的，这样的乡建作品实际上是为城市里的人所准备的。

这并不是否认为城里人服务的乡建的价值。即使是一种选择性的图像所提供的短暂抚慰也仍然是有益的。这里想要说明的是有必要将这种方式与另外一种乡建区分开来，那就是为乡村所做的乡建。贺勇在鄣吴所完成的就属于后一类，在这些项目中，乡村从背景中走向前台，并且发出不容拒绝的声音，从设计到使用，乡村始终占据着核心的位置。对于乡村来说，做别人的布景还是自己做主角，差别当然是明显的。布景随别人的剧情所摆布，可被替换也可被舍弃；主角不仅能掌控剧情，更重要的是还能不断拓展新的剧目。从车站到小卖店，鄣吴在建筑师的帮助下改造乡村环境的举动持续不断，而实实在在获益的则是村民。为乡村所做的乡建不是城市生活的补药，而是乡村生活的自主延伸，这实际上是乡村聚落演化转变的主要方式，而不是依赖于城市建筑师的“点石成金”。

在另一个角度看，为乡村而做的乡建也对城市人也有特殊的价值。在为城里人服务的乡建中，乡村被美化成对立于城市缺陷的“乌托邦”，但不应忘记，“乌托邦”并不存在，不去对身处的现实进行改变，“乌托邦”永远都是乌有之乡。这样的乡建，所能提供的帮助始终是有限的，而如果满足于此，我们甚至有可能错失真正改善的动机与机会。与之相反，由乡村主导的乡建始终是积极参与性的，恰恰是因为村民们不认为自己所处的是“乌托邦”，所以才需要不断的修正和改进。他们心目中也有一个理想的图景，并且愿意为这一图景付诸行动。罗伯托 · 昂格尔（Roberto Unger）将这种愿景称之为前瞻性思想 (visionary thought)，它“并不是完美主义或者乌托邦式的。它并不常常展现一幅完美社会的图像。但它却是要求我们有意识地重新绘制地图，来呈现可能的或者值得期待的人类关系，去

to the rural villages even though they are built in that area.

When the country villages became a mere background for the architecture projects instead of the actual initiator and users, the village will turn to what Adolf Loos described "Potemkin City", a superficially decorated illusion only for people to admire from afar[3]. Loos once used the same term, Potemkin City, to describe the city of Vienna in the early 20th century. People in Vienna built their houses in Renaissance style as if it could turn them into the nobles, when in fact they were not fooling anyone but themselves. Some rural projects are more or less doing the same, creating one after another "utopianized" rural village. "Utopia" is simply a fixed image silencing all the different voices because being perfect already.It allows no room for modification or suggestion. For some architects, the rural area is an "utopia". What it has—its natural conditions, lifestyle, and architecture features are something they couldn't find in city life. Therefore they see the rural projects as opportunities to make up for the defects of city life, providing comforts to people who have all sorts of complaints about urban life. Naturally, their works would be based on a selective description of rural life and on those rural elements which can fill in a gap for those missing in city life. What's left will inevitably fall into silence. This kind of designing work are actually for the urbanite, not for rural people.

I'm not trying to deny the value of the urbanite-oriented rural projects. In fact, I admit that even a fleeting comfort from them has its benefits. What I'm trying to say here is that it's necessary to make a distinction between the mainstream rural-building model to an alternative one, the one that's really for the rural life. He Yong's works at Zhangwu clearly belong to the latter category. He pushes the village itself from the background to the center of the stage, acknowledging its own unneglectable voice. From the perspective of the village the difference between being a mere backdrop and being a leading actor is certainly obvious. A backdrop could be changed, replaced or even discarded depending on how the plot goes, while being a leading actor, it can not only control the storyline, more importantly, it can constantly develop new plot for its play. From the bus station to the convenience store, Zhangwu are developing new renovation projects one by one, with the help of its architect. And the true beneficiary are the villagers. Rural renovation is not a complement for city life, but a development of its

发明人类关联的新模式，并且去设计体现它们的新实践安排”[4]。大卫・哈维 (David Harvey) 写道，这也就意味着一种辩证关系，“只有改变机制世界，我们才能改变自己，同时，只有基于改变自己的意愿，机构的改变才有可能”[4]。这种实践性的前瞻性思想与静态的乌托邦幻想之间的区别，也是为乡村服务和为城市服务的两种乡建之间根本立场的不同。它们导向的结果也不同，一种状况下乌托邦向空想领域越滑越远，而另一种情况下现实在辩证的改变中有可能越来越接近乌托邦。郭吴的案例具有很好的说服力，无论是建筑师还是村里，都没有一幅整体的理想图景，项目的推进也充满波折，但是在不断磨合与调整中，也还有垃圾处理站和小卖店这种更为成功的进展。如果一个村庄能完善地进行垃圾分类，有效地改善公共环境，城市的社区为何不能效仿，成为一个更为理想的“城市村庄”？这或许是来自郭吴镇的消息中最有价值的部分。这个措辞当然是在模拟威廉・莫里斯（William Morris）的小说《来自乌有乡的消息》。在那部小说里，莫里斯描绘了一个并不存在的乌托邦，来对现实进行批判，但是对于批判到乌托邦的道路却无人知晓。中国的乡村并不缺乏被“乌托邦”式计划所摆布的经历，而中国40年来的改革之路就起源于小岗村所发起的自我组织。郭吴镇的消息是关于前瞻性改进，关于建筑师与村庄的共同演出，关于在接受与改变中不断积累的经验与教训，关于出人意料的遗憾和出人意料的惊喜的讯息。这当然距离乌托邦的理想图景很远，但是与消息一同而来的是村庄一点一点地改变，乡建没有被定格于一两个项目，而是作为进程，不断到来。

在贺勇的这几个项目上，我们看到的是难以预知的发展过程，这提示我们对城市与乡村、经典与乡土、精英与日常之间的关系做出不同的思考。其实在郭吴村的传统中早已蕴含着促使我们颠覆这些二元划分的因素，这个村里所出产的竹扇，从选材、色泽、形态到结构无不精雕细琢，文人雅致耐人寻味。如果认同这样的精英制扇传统，又为何不能接受巴拉甘或者是阿尔托的

own right. It's not a stage for the magic show by the "golden finger" of the architect, but a way to evolve and transform for rural villages.

From another perspective, the rural construction has its special values to urbanites. When rural villages were beatified as some kind of "utopia" standing in stark contrast to the imperfect city life, urbanites need to remember that the so called "utopia" does not exist in the real world. Without making any actual changes to the reality, "utopia" can only be a "neverland". Such rural construction an provide limited help, and if we are satisfied with this, we may even miss the incentive and opportunity for real improvement. On the contrary, the rural-oriented building projects encourage all the active participation from the beginning. They are in constant rectification and improvements simply because villagers refuse to see their home as some kind of "utopia". They have their own vision of an idealized world in mind and they are willing to take actions to make it come true. Roberto Unger, the Brazilian philosopher, called this vision "visionary thought". According to him, "it's not perfectionist or utopian, nor is it a picture of a perfect society. However, it requires us to consciously re-draw the map, to present any potential or worthwhile human relationships, to invent a new model of human connections and to design new practical arrangement for them."[4] For that, David Harvey, the distinguished British anthropologist once commented it as a dialectic relation by saying, "only by change the working mechanism of the world can we change ourselves. Meanwhile, only by changing our own thoughts will it be possible to change the institution."[4] This distinction between the practical visionary thought and the static utopian dream also represents the opposition between urban-oriented and rural-oriented architecture. Obviously, they lead to different results too. The former gradually turns into a fantasy while the latter, in its constant dialectic changing, might be able to make the utopian dream come true. Zhangwu project can serve as a quite persuasive example. At the very beginning, neither the architect nor the village authority had a perfect and detailed plan, and yet as the projects progressed in constant negotiation and adjustment full of twists and turns, the outcomes turned out quite successful. The garbage disposal center and the convenience store are two cases in point. If a rural village can have such a sophisticated garbage recycling system to effectively improve its public environment, why can't a urban community emulate and turn itself into a even better "urban village"? This should be the most valuable part in "news from somewhere called Zhangwu", which could be seen as an parody of William Morris's novel News from Nowhere. In this novel, Morris depicted a non-existent perfect world called "Utopia" as a criticism to the real world. In China, the rural area has always been the designated stage for this kind of utopian plans to play out. For instance, China's 40 years of reform had started its journey from an ordinary rural village called "Xiao Gang". News from Zhangwu Town is about a visionary improvement, a beautiful play co-starred by an architect and a village and the ever accumulating lessons and experience from acceptance and alteration. Coming together with the news are the gradual change in the village. Rural construction isn't limited to one or two projects. It's a working process, ever changing, and ever arriving.

What we can see in He Yong's projects is the unpredictable development of this process, which is a constant reminder that we are supposed to rethink the relationship between cities and the country, classic and tradition, elite and ordinary. In fact, the exquisite cultural specialty of Zhangwu village has overthrown all the attempts to put a binary division among the above-mentioned concept. One of the special local products at Zhangwu is bamboo fans. They are pure works of art. From the material, color, design to structure, everything is so elegant and thought-provoking. If we can appreciate such an elite fan-making craftsmanship, why can't we accept the equally elite architectural styles like that of Barragan or Alto? Why do we have to draw an uncrossable dividing line between the East and the West or between elites and the country. When we talk about the country, we should also keep what Edward Said, a public intellectual teaching at Columbia University, has said in his criticism about orientalism in mind. What exactly do we want to see? A real rural village or an alienated village that's negatively defined based on a dichotomy ideology?6 A possible "ruralism" might be able to bring a similar damage. It could render people blind to a dynamic rural life and to its power " that can be truly touched and experienced".[6]

News from Zhangwu is not just a reminder of the necessity to reflect on our understanding of the rural areas.It can extend to other architectural discussions concerning the concepts in binary

村里正在夯土的工匠 I Craftsmen Tamping Earth in the Village

精英建筑传统？又为何一定要在东方与西方、精英与乡土之间画上不可逾越的分割线？当我们谈到乡村时，不应忘记爱德华·萨义德（Edward Said）对东方主义的批评，我们想要面对的是一个真实的乡村还是一个根据二元对立的需要“反向”（negative）定义出来的“异类”（alien）乡村？[5] 一种潜在的“乡村主义”可能带来的危害也是类似的，它会让人们忽视乡村中的能动性，忽视它“能真实感受到的，体验到的力量”。[6]

鄣吴镇的消息所提示的不仅是对我们理解乡村、切入乡村的反思。它也可以拓展到其他与常规的二元对立概念结构相关的建筑讨论中，比如传统、本体、阶层等等话题之中。无论是“尊重”还是“批判”，一种动态的、参与性的辩证互动，都比“乌托邦”布景更有利于建筑实践的可能性拓展。从这个角度看来，鄣吴村头车站的巴拉甘式厕所可以被视为一个标志，它的冲突与张力喻示了干涉与对抗，这也意味着拥有更多可能性的未来。

鄣吴仍然在践行这样的策略。在镇卫生院工地，我们看到建筑师仍然与甲方代表在现场讨论这里是否要增加一个房间，那里的影壁是否仍然需要。这又是一个典型的贺勇式乡建作品，或许不能再称之为乡建，因为它的体量已经扩大到数千平方米。经历过这么多剧情起伏，我们有浓厚的兴趣期待鄣吴镇所传来的新讯息。

（此文已发表于《建筑学报》2016年第8期）

opposition, such as the topics about tradition, noumenon, and social class, to name just a few. Whether it's "respectful" or "critical", a dynamic, and participative debate will always be beneficial for potential development and expansion of the practice of architecture. In this sense, the Barraganstyle toilet at the bus station at Zhangwu village can be considered as a symbol. The visible contradiction and tension on it speaks of the intervention and resistance that had happened to its construction. Meanwhile, they also heralds a future of more possibilities.

Zhangwu is still following this strategy. At the construction site of the town hospital, we saw the architect was in the middle of a discussion with the Party A representative about problems like whether an extra room was needed or whether that screen wall was actually necessary. This is another one of He Yong-style rural-building project. Except the scale of this one is particularly big involving a space of several thousands of square meters. We are looking forward to hearing the latest news coming from Zhangwu.

(This article has been published in the 8th issue of Architectural Journal in 2016)

References

[1] BEHNE A. The modern functional building [M]. Santa Monica, Calif.: Getty Research Institute for the History of Art and the Humanities, 1996: 123.
[2]ORTEGA Y GASSET J. Meditations on Quixote [M]. New York ; London: Norton, 1963: 46.
[3] LOOS A. Spoken into the void : collected essays 1897-1900 [M]. Cambridge: MIT Press, 1982: 95-96.
[4] HARVEY D. Spaces of Hope [M]. Edinburgh: Edinburgh University Press, 2000: 186.
[5] SAID E. Orientalism [M]. London: Penguin Books, 1995: 301.
[6] SAID E. Orientalism [M]. London: Penguin Books, 1995: 202.

村里的工匠师傅 | Craftsman of the Village

鄣吴十二舍

Studio 12F Office in Zhangwu

鄣吴镇鄣吴村 | Zhangwu Village，Zhangwu Town

建筑师：贺勇，孙姣姣，戚骁锋，纪敏，陈耀
建筑面积：274 m^2
结构形式：木结构（保留的主房），砖混结构（新建的餐厅）
建成时间：2017 年 7 月
Architects: He Yong, Sun Jiaojiao, Qi Xiaofeng, Ji Min, Chen Yao
Gross Floor Area: 274 m^2
Structure: Timber Structure（Reserved Main House）
Brick–concrete Structure（New-built Dining Room）
Date of Completion: July 2017

1. 主街 Site Location
2. 五凤巷 Wufeng Alley
3. 十二舍 Studio 12F Office
4. 穿村小溪 Stream

总平面图 Site Plan

引言

郭吴十二舍，其实只有一舍，原是一栋被主人弃用的老宅，位于浙江省鄣吴镇鄣吴村，周围群山环绕，翠竹繁茂。经过一番重新设计与改造之后，其作为十二楼建筑工作室在此工作和生活的场所，故命名为“鄣吴十二舍”。源于我们一贯的理念，不想谈太多关于空间、概念、文化等这些过于宽泛或宏大的话题，这次，我们想谈谈房子背后所在的村落、工匠、邻里以及生活，告诉你一个或许有点不太一样的建筑改造的故事。

壹 聚落

乡村之美，美的不是一栋栋孤立的房子，而是聚落的整体格局，鄣吴村就是这样。这里的绝大多数房子都是在最近二三十年以内建造的，但是其历史上由“八府九弄十二巷”以及穿村小溪构成的空间状态却相对完整地保留了下来，农居也维持着高密度的紧凑状态，在村里穿行，迷宫般的街巷总是会给你许多惊喜，你会发现很多自然随机的场所，也会碰到不少西扎式的偶发空间与形态——墙角里生长着野花，头顶上穿越着电线，不同年代建设的夯

Introduction

It's originally a deserted old house located in the village of Zhangwu, Zhejiang Province. The house is surrounded by rows of bamboos and the mountain. After undergoing a re-designing and renovation, it has been served as the living and working quarter for Architect Studio 12F Office in Zhangwu. As usual, we won't talk too much about any empty concepts like space, or culture. Instead, we will focus on the story behind the house, the craftsmen involved in its construction, its neighborhood and the rural life. We want to tell you a different tale about an architectural renovation.

1. The settlement

The true beauty of the rural village lies not in any single house, but in the whole complex pattern of the residence area. Zhangwu village has the same feature. Although most of the houses here are comparatively new, having been built within the recent 20 to 30 years, its old spatial layout of "eight mansions, nine lanes and twelve alleys" has been preserved completely. The houses are in a tight-knitted structure. Walking through the village feels like exploring a maze which constantly brings you surprises. You would encounter plenty of randomly-formed space and some Alvaro Siza-style spontaneous spatial forms and items, such as the wild flowers growing out at a corner or the foot of a wall, the traversing electric lines flying over your head. The old compacted mud blocks, grey bricks and

土、灰砖、白墙等各色房子混杂在一起，让村庄弥漫着时间的味道。或许理想的村子就应该是这样，一切似乎都在其适宜的状态，混中有序，杂而不乱。

郭吴十二舍（以下简称十二舍或老宅）就位于这样一个村子的边缘，加上地势较高，所以看起来与村子中其他部分有点疏离的样子，农用三轮车都不能通达那里，只能步行。或许是为了弥补这块场地在交通上的不足，两株高大的槐树守护在了这里，在盛夏里为这栋房子遮掩下一片浓浓的阴凉。

贰 五凤巷

因为老宅位于郭吴村的一个偏僻角落，到达那里，先得穿过一条不短不长的巷道。巷道的起点是一个小小的空地，其上原来有一栋房子，尽管已经被拆，但是在与其相邻的建筑上留下了一个大大的人字坡，还有斑驳的墙面，墙角歪歪斜斜地堆着一些大肚细脖子的陶罐。不知从哪个

whitewashed walls which were built in different years are all mixed up in houses of different colors, creating a sense of time. Perhaps this is what an ideal village is supposed to look like with everything staying in its right place in an orderly mess.

Studio 12F office in Zhangwu (the office or the old house for short) is located right on the edge of the village. Due to its slightly higher footing, it looks a little separated from the whole village. Agricutural tricycles are not accessible. We can only go there on foot. There are two giant honey locust trees standing near the house, casting a cool shade for it during the hot summer.

2. The alley of Wufeng

Since the old house is located in an outlaying corner away from the village, you have to walk along a moderately long alley to get there. The starting point of the alley is a tiny empty space with a big Y-shaped ramp. There are also several pottery pitchers, each with a big belly and a narrow neck, piling at the foot of the bounding wall of the next-door neighbor. A beam of sunlight beats down on the ground revealing its rugged and bumpy surface.

A stone door frame serves as the real starting point of

改造前的房子 | Original House

方向投来的一束阳光，漫不经心地洒在地上，地面的凹凸不平一览无余。

一个石头的门框，是巷道的真正起点。穿过去，才发现一条溪沟躲在它的后面。溪沟边总有人在那里洗衣洗菜，偶尔也会用棒槌敲敲打打。水里长了不少水草，很多小鱼在水草间穿来穿去。巷子两边都是人家，他们有着大大小小的院子，晒着五颜六色的衣服以及日常的生活物资。越往深处，巷子越窄，一户人家院子里的凌霄长得非常茂盛，攀出墙外，将花瓣撒满了一地。巷子尽端，是几十步由块石砌筑的台阶，拾级而上，便到达十二舍了。

叁　老屋 / 十二舍

老屋在改造之前，已经被原来的主人弃用多年，墙体残破倾斜，结构破乱不堪，已然危房。这栋房子其实并不古老，大约建于1980年前后，不过采用的依然是传统的木架构与夯土外墙，由主房、辅房和院子构成。主房里原来住着兄弟两户人家，其中一户两个开间，另一户三个。辅房是后来搭建的厨房，分为两半，每家一间。厕所建在另外的地方。建设之初，受制于有限的经济条件，木构架的用料都极其简陋，另外，由于居住空间的逼仄，主人也对这栋房子进行了多次的改造，那个三开间的人家在入口侧的坡屋顶下居然有着一部分混凝土的楼板，想来那部分原来应该是外廊，后来被主人扩展至房间之内。屋面一半小青瓦，另一半则是水泥瓦，如此混搭，倒也不觉奇怪。院子里的场地用水泥进行了硬化，一些植物依然顽强地从缝隙里生长了出来。

改造之初，应各方面的要求，就定下了主房大修、辅房原址重建的计划。新的功能定位为居住兼办公。在此项目中，我们自己也想看一看，在保持原本的外墙、结构体系以及形体的前提下，究竟在多大程度上可以优化内在的空间布局，提升整体的居住生活品质。新的布局中，将原来两户的空间整合到一起（但应房子原主人的要求，基本保留了两户之间原有的分隔墙体），根据当下的生活习惯，形成客厅（兼工作室）、卧室、卫生间、淋浴间、厨房等各功能房间。

辅房重建后，作为一个餐厅或讨论室，临近大树的那面墙开了一个大窗，可以看到窗外的村落和远处起伏的群山。客厅那部分的屋顶原本就矮一点，干脆通高，卧室部分屋顶略高，其阁楼正好可以作为三个小房间。功能调整

the alley. Stepping through it, you will find a little creek hidden behind the door frame. Occasionally, you can hear the beating sound of the wooden laundry stick owned by local villagers who come to the creek to do their laundry. There's quite a lot of water grass visible from under the surface of the stream with tiny fish swimming in and out of them. The alley is flanked by local residence with their own courtyards big and small, in which you can see clothes of various colors hanging on the clothesline. The alley become narrower as you walk along. One particular house boasts the most flourishing trumpet creeper, peeking out of the wall and casting its petals all over the place. At the end of the alley lies a short stone stair. Climbing up the stairs and you will be at the entrance of the office.

3. The old house/12F Office

Before the reconstruction, the old house has long been deserted by its former owner. The structure and walls were so dilapidated that the whole house looked shaky and ready to collapse at any time. Being built in the 1980s, technically, it's not an "ancient" house. It's just old-fashioned, adopting one of those traditional wooden framework and mud-brick walls. The house consists of three parts: the main house, side rooms and the courtyard. The main part which counts five rooms in total used to be occupied by two brothers and their family. There are two side rooms functioned as two kitchens, one for each family. The outhouse was built in the other part of the courtyard. It was said that back when the houses were built, the materials were all very simple and crude due to the owner's poor financial condition. In later years, because of the shortage of space, the two brothers had made multiple alterations to the building. Some of the changes are still visible today. For instance, there is a piece of concrete floor under one side of the roof at one of the entrances, which was supposed to be part of the porch, but clearly it had been expanded and became part of the indoor floor for another room. Another evidence is the mashup roofing tiles. Half of them are blue tiles and the other half are concrete tiles. Oddly, this mixed-up arrangement looks sort of okay. The surface of the courtyard had been hardened by cement. Nevertheless, some especially stubborn plants still managed to grow out of the crevice of the ground.

At the onset of the renovation project, it has already been decided that the main house is to undergo a major remodeling while the side rooms are to be reconstructed while remaining at the original location. New functions of the houses are to be for both living and working. From this project, we also want to find out that without altering the main form and structure, to what extent can the interior spatial layout be optimized and therefore the overall life quality could be improved. For the new layout, the two formerly divided living quarters will be integrated. (Though the original dividing wall between the two parts have remained intact just as how the original owner of the house wanted.) And following the modern day lifestyle, the new layout includes living

后的入口处，原本是一个外墙后退形成的凹空间，于是在这里增加了一个外廊，与餐厅挑出的外廊一起，既给建筑的形体添了些层次，也形成了几处半室外的场所。几把竹椅子放在那里，让人随时可以坐在那里看看院子里的风景。

肆　庭院

传统房子一般都是有院子的，院子和房子才构成了一个完整的居住单元。十二舍很幸运，有着一个比较宽敞的院子。主房朝着东南，所以房前的院子里每天都早早地洒满了阳光。因为地势较高，从这里可以看到村子里大部分高高低低、深深浅浅的屋顶。顺着院子外侧的边缘，原本就种有竹子一丛，枇杷、石榴、桂花各一株，竹子长得特别茂盛，一些爬藤植物顺着竹竿爬到了顶端，将其压弯了腰，耷拉在入口院门的上方。枇杷树已经被虫害严重侵蚀，枝叶稀疏、营养不良，不过其苗条舒朗的枝干，看起来倒也是玉树临风的样子。这个春天，工匠师傅在院子里

room, workshop, bedrooms, bathroom equipped with shower equipment and kitchen. As far as side rooms are concerned, since they are to be renovated as a cafeteria/conference room, a large window has been installed into the wall, through which you can see the whole village as well as the rolling landscape of the mountain in the distance. In addition, the formerly lower ceiling of the living room has been lifted up while the attic above the bedroom has been reshaped into a tiny room. There used to be a concave space created by the receding exterior wall at the entrance. After the renovation, a new porch has been added to that space and together with the porch outside the cafeteria, the newly added parts have given the whole building a sense of layers. By putting several bamboo chairs on the porch, it allows people to sit on the porch and admire the beautiful scenery of the courtyard anytime they want.

4. The courtyard

For a traditional Chinese residence, courtyard is an indispensable part of the whole unit. Luckily for 12F office, there was already a relative spacious courtyard to it. The main house faces the southeast which means the courtyard is

种植了两棵南瓜苗，如今，南瓜藤已经肆意地爬满了院墙，六七个南瓜在院墙顶端或角落安静茁壮地生长着。

院子里原来是水泥地，既不生态，也显得生硬，所以决定把它们砸掉后，重新铺上石板和瓦片，说起来轻松，这可是一件大工程。单说那水泥瓦片，尽管是屋顶上拆下来的旧物再利用，但因其尺寸太大，先得把它切割成三等分，用两头，去中间，然后再把它们一片片立在场地之上，最后在其缝隙间填满粗砂，如此操作，三个工人一天下来也就只能完成十来个平方米。到最后实在承受不起，加上水泥瓦被新的院墙用得差不多了，于是院子里的铺装改用了水泥砖，辅以少量的瓦片，如此下来，材质的肌理没能统一，不过倒也显示出因材而建的随性与自然。

伍　工匠

房子的改造当然是需要绘制图纸的，但是实施的过程与质量全在工匠师傅的手上。我们的图纸是非常粗糙的，一方面是因为对于木结构的传统工法知之甚少，只能画个大概，另一方面是老房子在改造过程中，在你掀开一片屋顶或在墙上凿一个洞之前，你根本无法知道会出现怎样的状况，往往只能顺势而为，现场设计。十二舍在改造过程中，很幸运地遇到了张师傅，他是本项目的施工负责人。张师傅早年学艺东阳，通晓老房子改造的每个环节，在他手上似乎没有解决不了的问题。张师傅家就在邻近的村里，手下所带领的一批工匠也都是本村或附近的，木工喻师傅的家就在十二舍的后面，只有一步之遥。大家一般早晨六点多就开工，中午十一点半回家吃饭，下午一点半再开始，然后五点收工回家。在张师傅的统领下，各个工种环节按时开始、按时结束，配合默契。有一天，木工骆师傅收工后开心地说，自己的任务完成了，明天要带着家人去西安旅游了。

在施工过程中，张师傅对于各项工种的要求很严格，其实他不在场的时候，各位师傅也依然能尽力地做好自己手头的工作。在楼梯的制作拼装过程中，为了踏步的平整，木工喻师傅和骆师傅反复地用水平仪调整自己的定位模具；切割瓦片的两位师傅，冒着高温和灰尘，动作干净利索，有条不紊；电工向师傅对于开关、插座的位置，比我思考得更加细致周全；院子门口的那块大石板，应该有好几百斤，为了把它放到合适的位置，三四个师傅合力一边调整，一边测量，汗如雨下，辛苦了半天，没有半句怨言。这些

flooded with sunlight every morning. Due to the high footing of the houses, standing in the courtyard and looking out at the village, you can see patches of rooftops high and low. Along the edge of the yard are a string of luxuriant bamboo trees intertwined with some heavy climbing plants, including a loquat plant, a pomegranate tree, and an Osmanthus tree. Serious erosion by pests, the loquat tree is already malnourished with sparse branches. But its slender and smooth branches also seem to have a special charm.This spring, two pumpkin seedlings were planted in the yard and now, their green vines have covered the whole wall among which six or seven little pumpkins are growing on the top or at the foot of the wall.

The surface of the courtyard used to be paved with cement, which was neither natural nor ecological, so we decided to replace the concrete with stone block and concrete tiles. This is easier said than done. Take the tiles as an example, since we are to re-use the tiles removed from the rooftop, some cutting and refining are needed before we could place them on the ground and fill out the gaps with rubbles and sand. The whole process is quite slow and exhausting. In the end, due to the lack of time and materials, we had to switch to concrete bricks. Though there is inconsistence in the materials, the finished work looks just as fine with a touch of spontaneity and naturalness.

5. The craftsmen

To a house renovation project, an engineering blueprint is undoubtedly indispensable. However, what's even more important is the highly-skilled and experienced craftsmen who are responsible for the actual construction. Our particular blueprint happened to be a bit rough with a lack of details, because the Chinese traditional carpentry work needed for the wooden structure of this house has been a rare specialty only known to the true professionals. Besides, the whole situation was quite unpredictable and in need of a lot on-site improvisation. Therefore, it's supposed to be especially demanding to construction workers. Luckily, we had Master Zhang working as the foreman of the construction team. Zhang started his apprenticeship in house-building in Dongyang at an early age. He knows everything about the restructuring of old houses like the back of his hand. It was said that no problem is unsolvable for him. Since Zhang and his crew all live very close to the construction site, they were able to get started as early as 6 o'clock in the morning and keep working till 5 o'clock in the afternoon except for a two-hour break at noon every day. Under Zhang's leadership and coordination, the construction site ran like clockwork. One day, after his work, Master Luo, another carpenter, said happily that he had finished his work and would take his family to Xi'an for a holiday.

During the construction, Zhang was quite strict with his crew. So much so that even in his absence, no one in his team slacked off in their tasks. For instance, in order to make sure each step on the stair was equally spaced and levelled off,

师傅们常说，“做工就是做生活”，“要对得起东家和这门手艺”，这份朴素的认知应该就是所谓的“工匠精神”吧。在工地现场，看着师傅们忙碌的状态，让人真切地感受到文化传承不是学者写几篇文章、政府恢复几座老房子就可以完成的事情，它是一个由“工匠”“东家”“做生活”等构成的生态系统，缺一不可。只有在有工可做、有匠人可寻、有足够的报酬可挣的时候，传统建筑或建造的文化才能够以一种自然、可持续的方式真正传承下去。

陆 邻里

十二舍偏安村落的一隅，看起来与周边的邻居似乎没什么瓜葛，但是随着工程的进展，那些隐形矛盾与复杂关系纷纷呈现出来。辅房重建后，邻近大树的山墙屋檐出挑了约60厘米，前面的邻居看到后坚决反对，因为大树下是她家的菜地。我说能否补偿你一些费用呢？邻居说这不是钱的事情，那我说你看屋檐可以出挑到哪个位置，邻居拿

Master Yu and Master Luo, both carpenters, made adjustment with the level instrument repeatedly. The two workers who were in charge of tile-cutting braved the stuffing dust and scorching heat while going about their job in a speedy and orderly manner. The electrician was so meticulous and thorough about each switch and outlet setting that he could put any designer to shame. To move a pretty huge stone slate weighing hundreds of kilograms to its designated position, 3 or 4 craftsmen had to make multiple measurements and adjustments just to make it right. Sweats ran down their faces like raindrops and yet no complaints were ever uttered. They always say that "doing your job" is a life-long commitment and that the effort you put into the job got to deserve what your boss has paid you and the craftsmanship you have been taught. The words are simple, but behind them, I believe, lies "the spirit of craftsmanship". Watching the whole construction team working makes you truly feel that maintaining the cultural continuity takes much more than a few articles from scholars or one or two government projects of restoring old houses. Rather, it's an ecological system comprised of craftsmen, the "boss" and "a life-long commitment". Only when all those elements are available can the traditional architectural heritage be handed down from

个杆子指在了约30厘米的地方，并让我保证屋檐的雨水不能排到她家的菜地之上。我说就这么定了。主房里原来没有卫生间，所以决定在西北侧增加两个。刚刚把地沟挖好，准备埋管子的时候，后面的那户邻居不干了，不允许继续施工。我问其原因，他说厕所放在这个位置有味道，会影响到他家。听到此语，我和工匠师傅们都很是惊讶，因为他的家距离十二舍至少有十米以上，而且他家门口还有约两米高的院墙。任凭我们怎么解释或尝试弥补，好像都无济于事，其实我们也慢慢明白，这不单是厕所的问题，而是邻里间多年的积怨在这个节点上的一次爆发。这件事情僵持了很久，最后在村干部的协调下总算平息，施工得以继续。

化粪池的事情，不得不提。由于村里还没有污水集中处理的市政管线，所以需要自己建一个化粪池。从功能的合理性来讲，最理想的位置是餐厅东北角的三角形坡地之上，那里离十二舍的厕所距离短、位置低。经过多方打听，终于搞清楚那块地的主人，他早已经搬到新村居住，三角地对他而言无甚用处了。与他讲明情况，他非常礼貌地拒绝了，原因是不想与曾经的老邻居在将来有潜在的纠纷。无奈之下，那就决定埋在院子里吧。请一个微型挖土机刚刚挖好了一个大坑，结果房东自己来抗议了……最后，化粪池被迫埋在了院子外面南侧的竹林里，那里是原来房东家的地，与周边各邻居都相距甚远，只是这里离十二舍的厕所也相距遥远，而且管线绕了几个大弯，这意味着走管线的沟要挖得更深，管子也更容易堵塞。不过，张师傅也自有应对的办法，在管子的源头留足了通气孔，在管子的中段又接入了部分雨水管，保持对排污管线的经常冲刷，从而使其通畅。这么一番折腾之后，十二舍改造工期和费用都成倍增加。经过这些事情，我们发现村里的很多东西，表面上看起来似乎模棱两可、无人在乎，可事实上每块地、每棵树都有其主。平常这种权属关系都是隐藏着的，但是村民之间却都心知肚明。所以在村里造房子不仅是简单的一项建造工程，也是邻里间社会关系重塑的过程。不过，也正是在这种相互的博弈与约束之中，乡村肌理与风貌得以相对稳定地维持。

generation to generation in a natural and sustainable way.

6. The neighbors

The position of the 12F office is at a far corner of the village. It seemed remote and unlikely to get in the way of anybody. However, as the project proceeded, lots of seemingly nonexistent contradictions and complicated connections soon emerged. After the re-modeling of the side rooms, one of the neighbors immediately complained about the widen 60 centimeters of the breath of the eaves on top of the gable adjacent to a giant tree since it blocked the sunlight from her vegetable plot. I suggested that we could pay her a sum of money as a compensation. She refused and repeatedly explained to me that it's not a matter of money. Eventually, we both made some compromises and got the problem solved. Of course, that involved a little modification to the house setting on my part which required to add two toilet rooms to the northwestern side of the house. Just when we were on our way to dig out the trench for the pipeline-installing, the neighbor on the other side of the house came up and stopped the process. His complaint concerned the position of the toilet. He claimed that the bad smell from the toilet would reach his house and that would be a problem to his family. My team and I were completely incredulous and confused considering his house was at least ten meters away. Besides, the bounding wall around his courtyard was as high as two meters which would definitely block out any unpleasant odor. All the explanations or the monetary compensation that we came up with ended up in vain. In the end, we realized that it's actually not about the toilet, but a outbreak of the accumulated conflicts and grudges among the neighbors over years. The deadlock persisted for quite a while before problems got solved with the intervention from the village council.

One more thing that needs to be mentioned is the building of the septic tank. Due to the lack of a municipal pipeline system for centralized sewage treatment in the village, we needed to build our own private septic tank. An ideal location for it would be a triangle sloping field at the corner to the northeast of the cafeteria since it's close to the toilet and on a lower footing. That part of the land actually belonged to someone else who, we found out, had moved away to his new house located in another village and therefore would have no use of the sloping field. We contacted him and explained our plan. Unfortunately, he said no, with politeness of course. His reason was that he didn't want to have any potential disputes over the land with his old neighbors. Therefore, there's nothing else we could do but change the location to one within our courtyard. Hardly had we finished digging out the pit for installing the tank when the third complainer showed up and expressed his protest. This time it was the landlord himself. Anyhow, after so much twists and turns, the septic tank ended up in the bamboo patch outside the courtyard. The exact location was on the south side of the

柒　生活

经过大半年的施工，曾经濒危的老屋以十二舍的全新面貌重新出现在村里，也正式作为十二楼建筑工作室的现场办公和居住的场所。每当我和我的学生们在街巷里进进出出的时候，许多村民都会站在那里默默地打量我们，一些人也会直接进到我们的院子里，把房子的上上下下看个遍，然后转身离去。附近的一个聋哑人经常来这里光顾，每次他脸上都带着孩子般亲切的笑容，咿咿呀呀中，我不太懂他在表达什么，但我能看出他对这个房子的喜欢。住在县城附近的马经理在看了餐厅后，当即在自家的老宅里加建了一个同样的餐厅；前几天来安装窗帘的师傅说，如果他家的老房子还在，也要如此改造一番。

住在村子里，要买菜做饭、交水电费，要请人来安装网络、调整入户电线，要向邻居借一些东西……于是，在这日常生活中，对于村里的了解也渐渐多了起来，知晓了更多一些这家和那家的关系，谁家的孩子在哪里读大学，街角杂货铺的老板来自哪个村。前天在村里老街上走的时候，喻师傅从一个房子的屋顶上大声地叫我，原来他在帮那户人家翻建房屋。昨晚散步的时候，前面有两个人快速地走过来，走近一看，原来是昌硕故居边做竹扇的李师傅和他的爱人，我们彼此点头微笑了一下。

院子的一段围墙太矮，需要点植物攀爬在上面，突然想到鄣吴溪边原来种了很多凌霄，最近那里在改造，都被挖掉了，或许能捡一株过来。一打听，原来早被玉华村全部拿走了，于是去找他们要了两株种在了院子里，很期待再过两年它们会长成什么样子。

bamboos which was not exactly an ideal choice, because it's far away from the toilet which means the pipeline had to be especially winding and deep from the surface which led to a higher possibility of blockage. Fortunate for us, Master Zhang can easily solve the problem. At the starting end of the pipe, he installed plenty of ventilation holes and down the middle section he inserted rainwater downpipe to make sure there would be frequent scourings to the pipe to make sure it's unobstructed. However, all those extra works also cost extra money and time which almost doubled our original budget. An important lesson we had learned from those conflicts and negotiations is that every one of those seemingly derelict land, trees and plants actually have an owner of their own. There's a hidden social contract among the villagers to mutually recognize and respect each other's property and avoid any transgression. It may have posed a variety of challenges and obstacles to our renovation projects but those mutual restrictions do serve as a stabilizer for the working mechanism of the rural areas.

7. Life

After more than half a year's remodeling and construction, an old house presented itself in a completely new appearance with its new role as the living and working place for Architect Studio 12F. Every time when my students and I walked through the streets and alleys, quite a few villagers would stand nearby and watch us silently. Some of them would even walk into our courtyard and take a closer look at the house before turning around and leaving. Among those frequent visitors, one of them was a bit special. He was deaf-mute, but whenever I saw him in our courtyard, I could see child-like smiles on his face while making some yi-yi-ya-ya sounds. I didn't know what exactly he was trying to say but I could clearly see that he really liked this house. Mr. Ma, a restaurant owner admired the cafeteria so much he immediately had a new cafeteria of the same style built on his property. The guy who installed our window curtains even talked about remodeling his old house in the same way if possible.

Living in the village means you will get familiar with the villagers through everyday interaction with them such as shopping at the farm market, paying the utility bills, having people set up the electricity and Internet connection, and borrowing stuff from the neighbors. More and more we feel like we have become official members of this particular community where every family has its own story. You could never walk down the streets without having a chance meeting with someone you know.

The bounding wall around our courtyard looks too short and plain, so we decided to put some climbing plants for decoration. One choice is the trumpet creeper growing on the bank of a brook in Zhangwu. We managed to track down the disposed plants and moved two pots into our courtyard. Now we are looking forward to seeing how they would turn out in the coming years.

改造前平面图
Original Plan

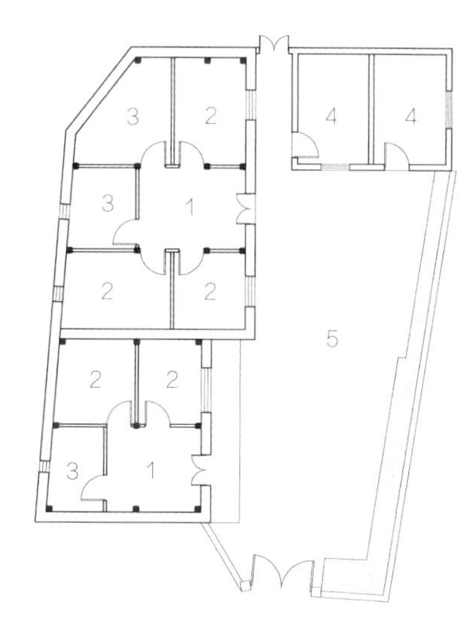

1. 堂屋 Living Room
2. 卧室 Bedroom
3. 储藏室 Storage Room
4. 厨房 Kitchen
5. 院子 Courtyard

一层平面图
First Floor Plan

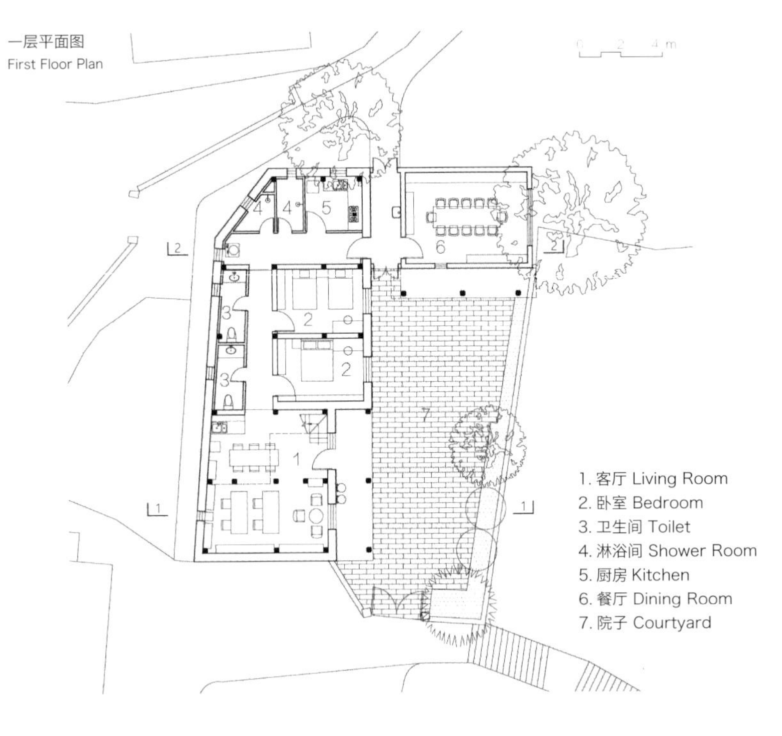

二层平面图
Second Floor Plan

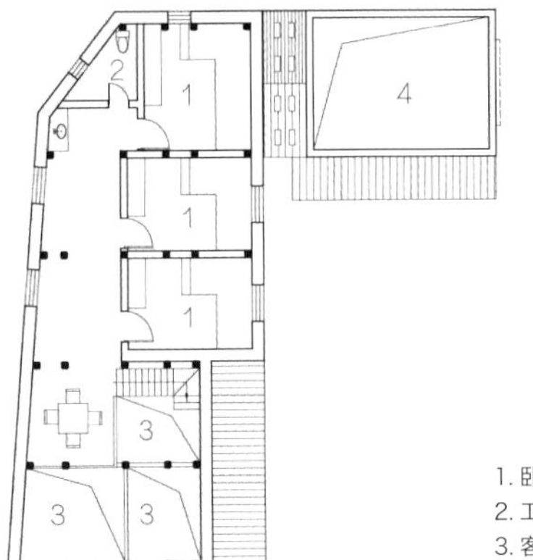

剖面 1-1
Section 1-1

剖面 2-2
Section 2-2

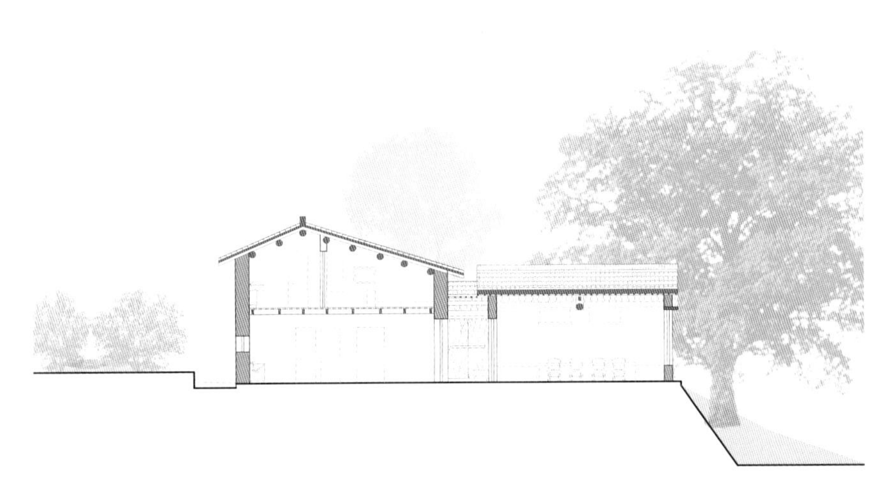

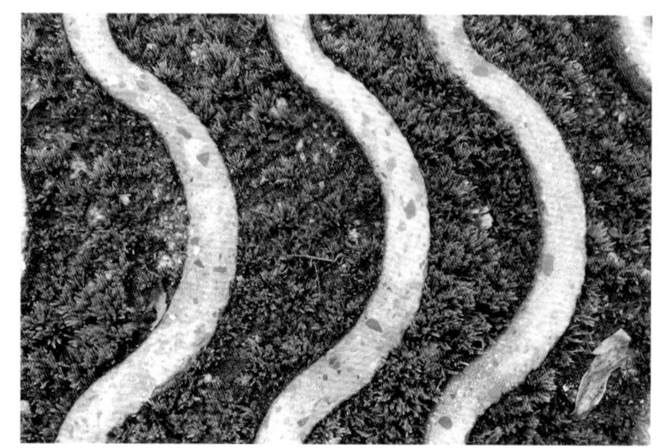

景坞村旅游接待中心

Tourist Center of Jingwu Village

鄣吴镇景坞村 | Jingwu Village, Zhangwu Town

建筑师：贺勇、倪书雯
建筑面积：265 m^2
结构形式：混凝土框架结构
建成时间：2013 年 3 月
Architects: He Yong, Ni Shuwen
Gross Floor Area: 265 m^2
Structure: Concrete Frame Structure
Date of Completion: March 2013

总平面图 Site Plan

该建筑位于村中心的一个三角地，周边均被道路环绕。设计之初，定位为旅游接待中心，兼具微型购物中心功能，我们设计了两组由几个基本的矩形空间单元（每个单元的面积为4.2 m×6.9 m）彼此连接而成的形体，单元之间可分可合，以满足未来各种灵活的用途，另外，用一个长廊将这组房子与外部的场地连为一体，形成房子、院子、连廊等多种有趣的空间状态。可事实证明，这只是我们一厢情愿的想法。建成之后，长廊被完全抛弃，干净的“白盒子”的周边也被增加了一个由深色涂料勾出的框架。不过，被村里调整过的房子却与场地的环境以及周边的房子更好地融合在了一起，给人感觉它似乎原本就存在于那个地方，而不是建筑师刻意设计的结果。现在，村里在房子中设置了甜品店、特色商品小卖部、图书阅览室，也举办过书画展，尽管都不如城里那般精致，看起来漫不经心，可似乎正好应和着村里的生活惯性与节奏。

This building is situated on a triangle-shaped land at the center of the village, surrounded by highways. At the beginning of the project, it was designed to be a tourist center with extra functions such as a mini-shopping center. Based on our plan, there will be two sections, each of which consists of several rectangular-shaped space units (with a square measure of 4.2 m×6.9 m) connected with each other. The units can be either separated or connected easily to cater for the requirement of any possible usages in the future. In addition, the houses are linked with the peripheral area by a long corridor, creating a string of intriguing space occupied by houses, courtyard, and hallways. However, it turned out that this is just a one-sided wishful thinking. After the construction was finished, the long corridor has been completely deserted. The clean "white boxes" have been painted by someone with dark color on the edges. Fortunately, the adjusted houses have perfectly merged into their surroundings, making people believe that they have always been there, not something that's been imposed on that place by the architect. Now, these houses have been fully used in the form of a dessert shop, a specialty boutique, and a small library which is capable of hosting book painting and calligraphy exhibitions. They are not as fancy as those in the city but they fit perfectly with the rural life style.

一层平面图
First Floor Plan

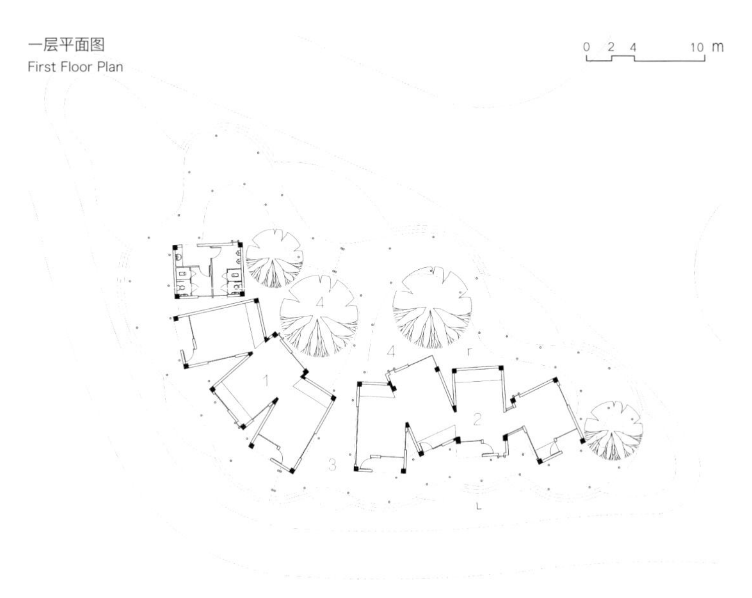

1. 图书馆 Library　2. 特色商品小卖店 Specialty Boutique　3. 连廊 Veranda　4. 院子 Courtyard

轴测图
Axonometric Drawing

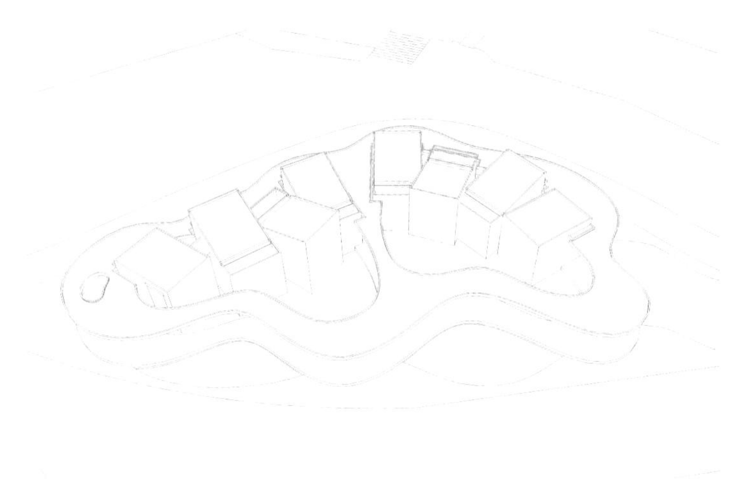

鄣吴村书画馆

Calligraphy and Painting Museum of Zhangwu Village

鄣吴镇鄣吴村 I Zhangwu Village，Zhangwu Town

建筑师：贺勇，朱博
建筑面积：465 m^2
结构形式：混凝土框架＋砖混结构
建成时间：2013 年 6 月
Architects: He Yong, Zhu Bo
Gross Floor Area: 465 m^2
Structure: Concrete Frame + Brick-concrete Structure
Date of Completion: June 2013

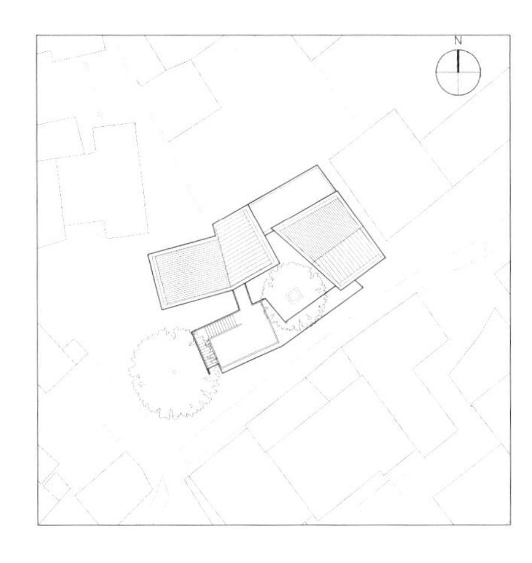

总平面图 Site Plan

该建筑位于村中心，原来是一处有着围墙的私人住宅，住户搬迁之后，该宅基地被收归为村集体所有。为了完善该村的休闲、旅游、文化设施，村里决定在此处修建一个书画馆，以收藏、展示本地一位书法家的作品，也以此作为一个供艺术家、村民、游客等休憩、交流的场所。

设计过程中，建筑师思考的首要问题是在一个狭小的场地上，如何将一个原来封闭的私人宅院转变成一个开放、公共的空间。第一件事情是去掉围墙，然后将不大的体量根据功能进一步打散，各体块的外轮廓顺着用地边界起伏转折，呈现出乡村那种随机、不规则的形态，体块之间也自然地形成多种空间的状态。从村内街巷中看过来，建成后的书画馆似乎只是一栋高一点的民居，白墙、灰瓦、坡屋顶，没有什么特别的地方，但是走近之后，却发现它的不同之处：这里有连廊、合院、街巷、小广场等多种公共空间的形态与相关设施，会让人们自然停留下来。

茶室顶部是开放的露台，通过室外台阶可以到达，这

This building is located at the center of the village. It used to be a private house enclosed in a bounding wall. After its owner moved away, this place has become a public property of the whole village. To add more entertaining and cultural facilities, the village council decided to build an art museum as a place to collect, preserve and exhibit the works of a certain local calligrapher. It could also serve as a public communication center for artists, tourists and local villagers.

For its design, the first question the architect had to answer was how to transform an enclosed and private residence to a public and open space. Apparently, the first thing to do is tear down and remove the bounding wall before breaking the building down into different small sections based on their different functions. The contour of each section is designed to fit with the winding and undulation of the dividing line among them, showcasing a series of patterns of random and irregularity. Walking through the lanes and alleys in the village, the art museum you can see looks just like a slightly taller residence house with white walls, grey tiles and hip roof, nothing special. And yet, coming closer, you will find something unusual about it. There is a wide variety of spatial patterns

也为书画馆的参观提供了多种路径。一棵原来保留的香樟，一株新种植的银杏，将建筑掩映在光影之中。

and facilities, such as vestibule, inner courtyard, small lanes, mini-square which make you feel like staying for a while.

There's also a tea room, on top of which is a balcony accessible from an outdoor stair which offers an alternative route for visitors to walk around the museum. Two giant trees, a camphor tree and a gingko, stand in the courtyard whose branches and leaves wrap the whole building in a bundle of lights and shadow.

一层平面图
First Floor Plan

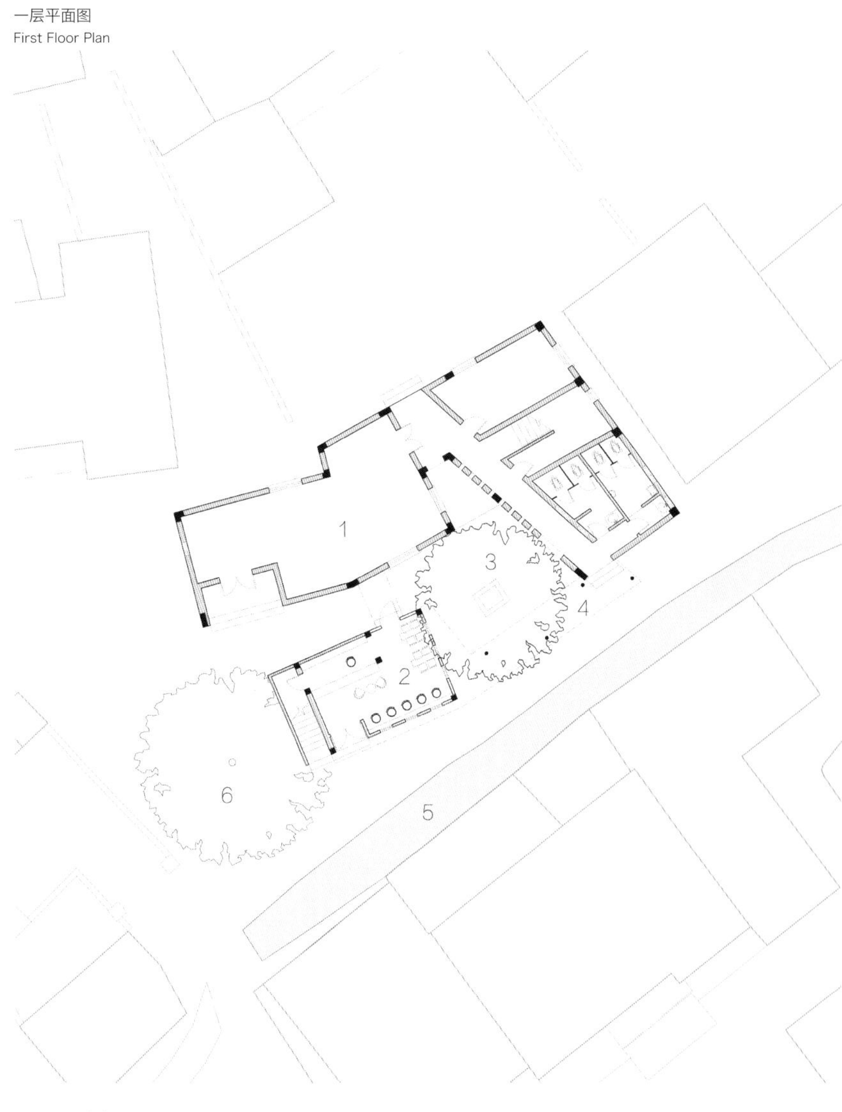

1. 书画展览室 Calligraphy and Painting Exhibition Room
2. 茶室 Tea Room
3. 内院 Inner Courtyard
4. 连廊 Corridor
5. 溪沟 Gulley
6. 小广场 Small Square

无蚊村小卖店

Little Convenience Store of Wuwen Village

鄣吴镇无蚊村 | Wuwen Village, Zhangwu Town

建筑师：贺勇，王竹，王静
建筑面积：78 m^2
结构形式：混凝土结构
建成时间：2013 年 11 月
Architects: He Yong, Wang Zhu, Wang Jing
Gross Floor Area: 78 m^2
Structure: Concrete Structure
Date of Completion: November 2013

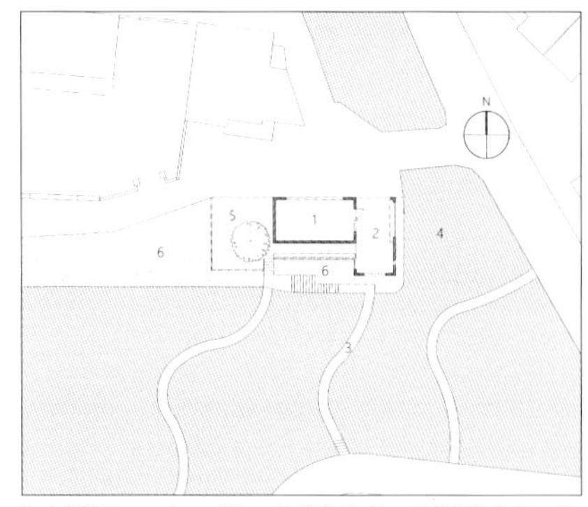

1. 小卖部 Convenience Store 2. 凉亭 Pavilion 3. 滚水坝 Rolling Dam
4. 水面 Water 5. 小广场 Small Square 6. 绿地 Green Land

总平面图 Site Plan

该建筑位于该村中心名叫“月亮湾”的位置（无蚊村是景坞村的一个自然村），这里几条溪沟在此汇合。在改造之前，河道里满是滚石、淤泥，景观杂乱不堪。临水边，原有一处村民自建的临时简易棚屋，是一个小卖店，出售一些日常生活用品，在美丽乡村环境整治中被拆除。建设过程之中有两种选择：将场地恢复成绿地景观，或者重新建设一个小卖店，但是产权归村里所有。最后与各方交流意见后，决定选择第二种方式，因为从村民日常生活的角度，需要这样一个商业设施。规划设计中，我们首先对三岔口原有的河道景观进行了设计，依据地形起伏修筑了3个低矮的滚水坝，滚水坝由平整的石块砌筑。水小的时候人们能在坝面上行走、浣洗。建筑没有采用通常的“乡土”或“传统”风貌，而是一个非常简洁、现代的清水混凝土的“盒子”，平面呈 L 形，长边布置售卖功能，短边则既是小卖部的前厅，也是一个供村里人休息、聚会、谈天的亭子。亭子的一面是开敞的临水靠椅，另一面则是月亮形圆窗，倒映在滚水坝常年蓄水的平静水面上。该建筑的建造过程也颇特别，工匠将此地山上的毛竹一分为二，钉在木板上构成模板，按照墙的位置把模板固定后，将拌好的混凝土注入其中，待其凝固、硬化后，拆掉模板，建筑便完成了。小卖店 + 凉亭的组合模式，成为触发该村乡村生活的“触

This building is located at the center of a village named Wuwen, which is a tiny hamlet under the jurisdiction of Jingwu village. The particular area is called "Moon Bay" because of the moon-shaped river bend where several creeks and streams converge. Before the re-construction, the river bed was full of cobble stones and silted mud, which looked extremely messy. On the bank of the river there used to be a small make-shift shack built by some villagers who turned it into a store selling some daily goods. Due to the fact that the building was unauthorized and a sort of damage to the natural scenery, the government had ordered it being demolished. There were two options for the renovation of this particular spot, either turning it to a part of greenland landscape or building a new store whose property ownership will belong to the village council. After some discussion among the concerned parties, the second option has been chosen. One of the reasons is that the villagers living nearby would benefit a lot from a grocery store in their daily life. Before the store re-building started, we made some modification to the scenic view of the stream channel by constructing three short rolling dams based on the natural undulating terrain. The dams are paved with stone blocks which are smooth and uniform in size. Whenever the water level receded, the surface of the dams would emerge for people to walk on or wash clothes. The new store looks like a box of fair-faced concrete, which is actually a simple and modern style. The layout of the store is in a "L" shape. The longer side of "L" is the main body of the store while the shorter side serves both as the entrance to the store and a

媒”和“针灸”，使得该设施建成后被村民们高效使用与自主管理。村民自己搬来了桌椅板凳，添置了麻将桌、儿童活动设施等。每天茶余饭后，很多村民、特别是小孩子们来此聚集、嬉戏，使这里成为全村最具活力的场所。随着时间的推移，旁边种植的竹子、树木渐渐长大，爬藤植物也慢慢上了墙，建筑与周遭环境很好地融合在一起。建设前的小店只是在空间上位于村里的中心，建设后的小卖店却成为村民日常生活的中心。

pavilion for villagers to gather together for a chit-chat. On one side of the pavilion sits back-rest chairs facing the river bend while on the other side you can see a moon-shaped window, reflecting on the calm water surface. This part of the building was constructed in a somewhat unusual way. First, the builders split the bamboo poles in halves and then had them nailed to a wooden board. They used these wooden boards as casting moulds and placed the moulds where the wall was supposed to be before filling them up with liquid cement. Once the cement solidified, the moulds were removed and there you have it—the building was done. This combination of a convenience store and a pavilion has become something of a "catalyst" which triggered a flourishing public life in the rural area. Ever since the building was finished and put to use, people living in the neighborhood started to treat it as a public gathering place. They gradually added more furniture and entertaining equipment to the pavilion such as a mahjong table, stools and even toys for children. Before long, this tiny pavilion has become the most active and vibrant spot of the whole village. As time goes by, the surrounding bamboos and other trees are growing taller and vines are climbing on the wall, covering its surface more and more. Gradually but steadily, the whole building merges into the environment seamlessly. Before the renovation, the store was at the center of the village only in terms of spatial position, and yet after the construction, it has become the heart of the village life.

场地原状 I Original Site

无蚊村小卖店

雪花
SNOW

雪花
SNOW

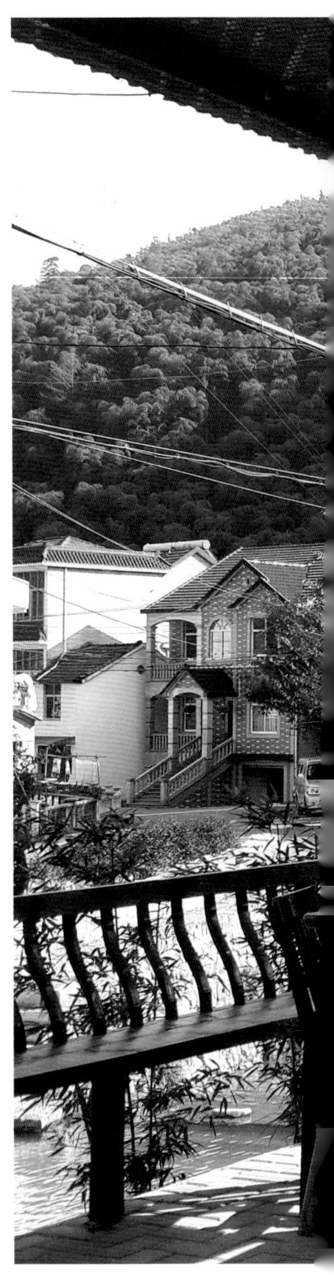

无蚊村小卖店

鄣吴镇垃圾处理站

Garbage Disposal Station of Zhangwu Town

安吉县鄣吴镇 | Zhangwu Town, Anji County

建筑师： 贺勇，金通
建筑面积： 184 m^2
结构形式： 混凝土框架结构
建成时间： 2014 年 10 月
Architects: He Yong, Jin Tong
Gross Floor Area: 184 m^2
Structure: Concrete Frame Structure
Date of Completion: October 2014

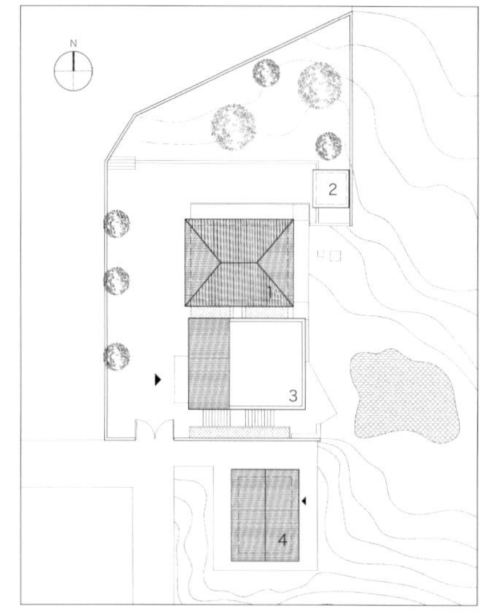

1. 原有站房 Original Station 2. 原有配电房 Original Power Room
3. 新建站房 New-built Station
4. 新建管理用房 New-built Management Room

总平面图 Site Plan

该设计项目的任务是在原有的垃圾压缩用房（那个白色四坡顶房子）的旁边，增加一个厨余垃圾就地资源化处置的房子。垃圾处理站的形体与建造几乎完全是一个追随功能要求与场地特征的结果：四坡屋顶形式源于垃圾车倾倒垃圾的需要；多孔水泥矿渣砌块的选择是通风的要求；建筑前面的几个水池是处理污水的湿地；利用场地高差，将处理设备安放在负一层，这样垃圾在地面层分类之后，不可资源化处置的垃圾则通过一个传送带运到旁边的压缩用房，定期运到填埋场；可处理的厨余垃圾则通过一个孔洞和管道，直接投放到负一层的处理设备之中；处理后生产出的肥料用三轮车通过坡道运出。

一切看起来简单，自然；在这份简单、自然之中，光线在室内也投下意想不到的动人效果，或斑斑点点，或倾泻而下，给令人作呕的垃圾堆放、分拣场所平添了些许愉悦。场地部分的室外红砖平台与台阶，当初并没有提供详细的图纸，建筑师只是在现场和工匠进行了讨论，具体则

The task for this designing project was to add a new room to the original garbage compressor plant (i.e. that white house with hipped roof) as a processing center for kitchen waste. The form and structure of the new garbage disposal station is the result of the attempt to pursue a perfect integration of functions and the site features: The hip roof design is to cater for the dumping operation of garbage trucks; the use of porous cement-slag blocks for the wall building is for the purpose of ventilation. The several water pools located in front of the building are turned to the marsh for sewage water processing; the height difference between two houses have also been made good use of–we put the processing machine on the minus one floor which is below the surface in that the non-degradable and non-recyclable garbage sorted out on the surface floor would be transported to the compressing room next to it through a conveyer band, from which the garbage will be on its way to the landfill. Meanwhile, the recyclable kitchen waste would be directly dumped into the processor downstairs where they will be reprocessed and turned into fertilizer before being carried out of the station by a three-wheeled mini-truck.

Everything seems so simple and natural.The seeming

由工匠们根据场地高差来灵活决定。事实证明，工匠还是具有非常强的“工匠精神”的，不仅合理解决了高差，还做出了部分红砖在墙面上出挑的细部。最终效果超出建筑师当初的期待。

simplicity and naturality has also been contributed by the unexpected illuminating effect from the indoor lighting. The streaming or speckled lights has actually created a somewhat pleasant ambience for the normally obnoxious garbage sorting place. The red brick platform and steps outside the house are actually from the pure improvisation and collective decisions of the builders after a brief discussion with the architect. Their meticulous and flexible handling of the details has showed "the spirit of craftsmanship" and a strong work ethic of our building-team. The final outcome was way beyond the architect's expectation.

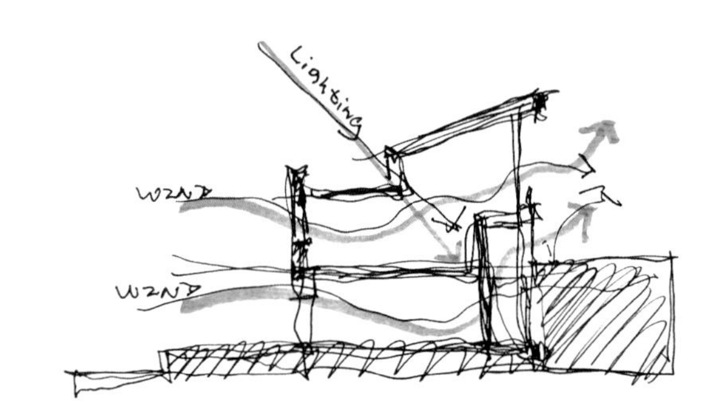

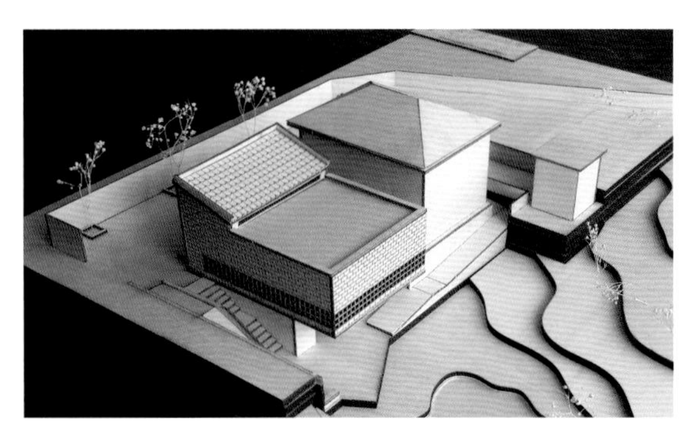

鄣吴镇垃圾处理站

一层平面图

First Floor Plan

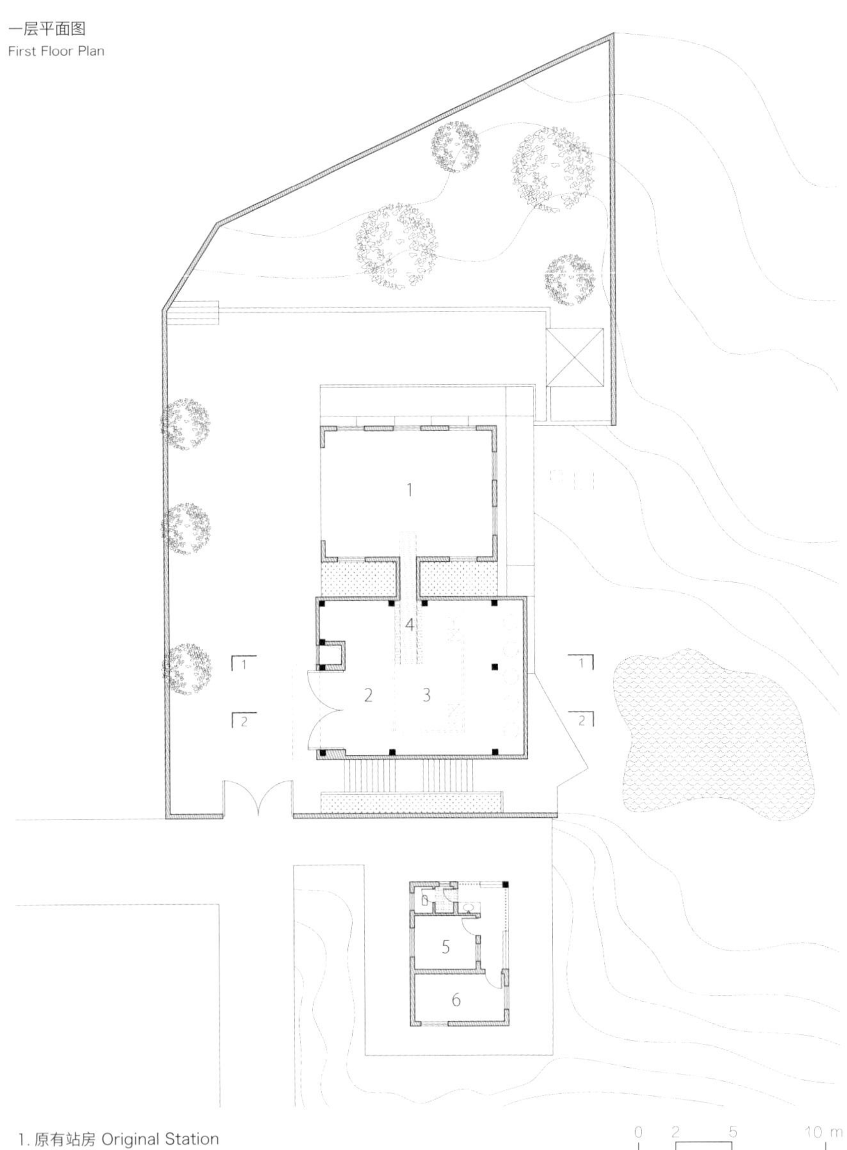

1. 原有站房 Original Station
2. 新建站房 New-built Station
3. 垃圾倾倒处 Dumping Area
4. 传送带 Conveyer Belt
5. 休息室 Rest Room
6. 管理用房 Management Room

有机垃圾处理车间

有机垃圾处理车间

剖面图 1-1
Section 1-1

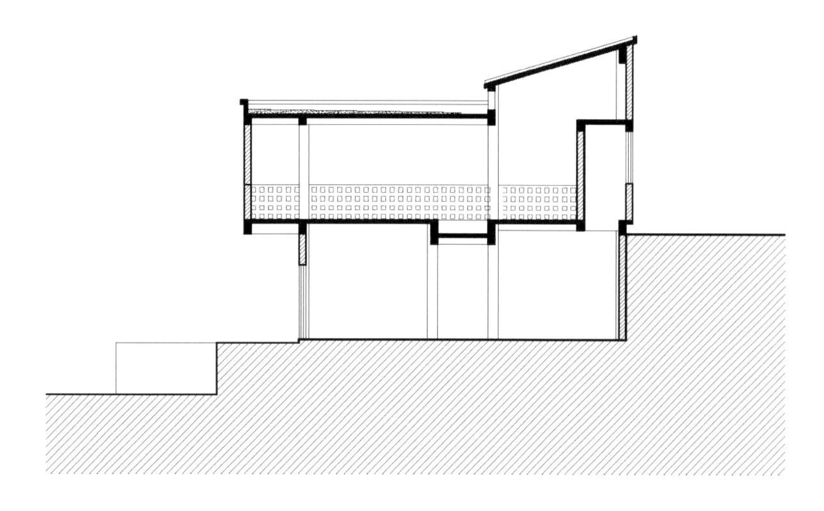

剖面 2-2
Section 2-2

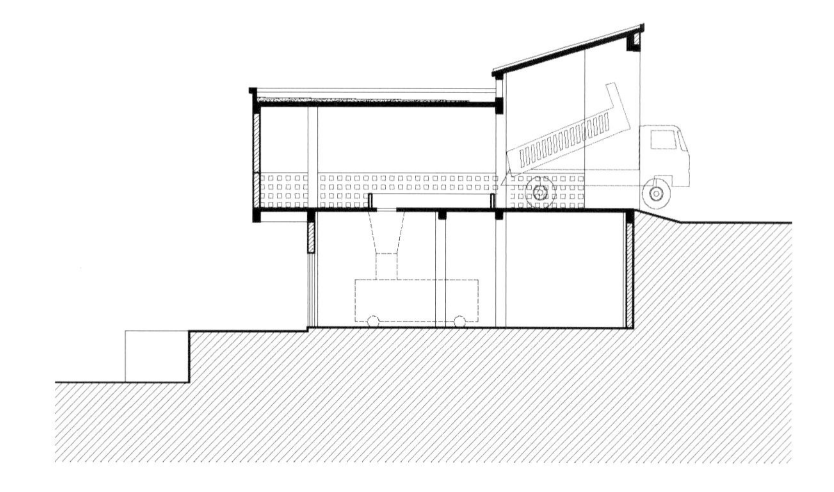

分拣车间

垃圾

建造过程 | Building Process

上吴村蔬菜采摘设施用房

Vegetable Planting and Picking Facilities of Shangwu Village

安吉县鄣吴镇丨 Zhangwu Town, Anji County

建筑师： 贺勇，曾伊凡
建筑面积： 213 m^2
结构形式： 砖混结构 + 钢结构
建成时间： 2015 年 10 月
Architects: He Yong, Zeng Yifan
Gross Floor Area: 213 m^2
Structure: Brick-concrete Structure + Steel Structure
Date of Completion: October 2015

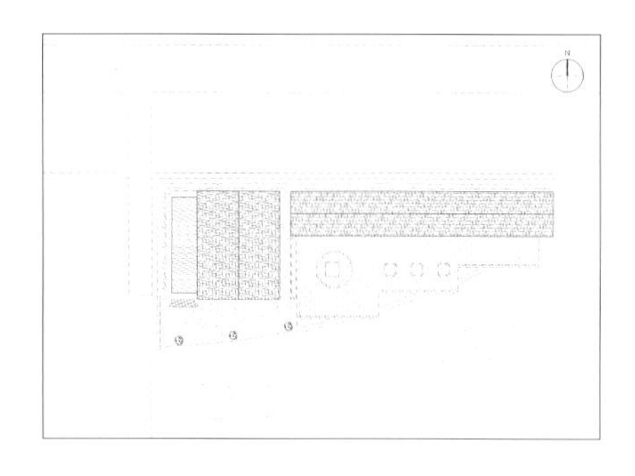

总平面图 Site Plan

在上吴村的一片农庄地里，出于蔬菜采摘、储存以及供游客喝茶休息的目的，有了建个房子的需要。关于建筑的布局与大致感觉，村里早已经有了想法，建筑师只不过是利用自己的专业知识，将其更加清晰、合理地表达出来。建筑采用了乡土民居的风格，并使其尽可能水平地伸展于场地之上。这样做的目的是尽量减小建筑对于原本极好的自然生态环境的干扰；或者说建筑的介入，只是为了让环境更好地呈现。所以，建筑的布局与形式完全无需多余的变化与装饰。一个封闭的仓库式蔬菜储存用房、一个尽可能开敞的连廊、一块干净的露台几乎构成了建筑的全部。山水与土地，决定了建筑的基调与尺度。于是，竹林旁、茶树间、田野中、池塘边，建筑静静地立在那里，可是那些缤纷的油菜花、绿油油的蔬菜以及劳作的人们，却丰满地呈现出一个从容自足的生活世界。

劳动力的缺乏与昂贵是这个项目中遇到的一个很大问题。这个项目中的工人基本都是村里的村民，如今村里的

To provide a place particularly for vegetable picking, storage and for visitors to enjoy a tea-break, a house is needed in a vegetable field of Shangwu village. The village council already has a rough idea about its structure and style, so what's left for the architect to do is clarify and refine this idea by putting in concrete details and make it come true using his specialty and professional knowledge. The style of this house is in line with that of rural residence. In order to minimize any possible disturbance from the building to the surrounding eco-system, there's no need to add any extra ornamentation or variation. The whole building is comprised of an enclosed storage room for vegetables, an opened-up corridor, and a plain veranda, nothing more. Landscape and land determine the tone and scale of the building. Therefore, it stands quietly beside the bamboos, between the tea trees, in the fields, beside the pond. It stands there silently against the backdrop of green vegetables, yellow flowers and the hard-working farmers, presenting a life of peace and self-reliance.

A shortage and the surprisingly high cost of labor was a big problem for this project. The craftsmen in this project are basically the villagers of this village. In average, a senior

一个大工（泥工、木工等）一天的费用接近300元，小工也至少要160元，因此如何快速地建造成为一个主要问题。于是，原本的夯土墙改为了空心砖加外抹黄泥，原本的木结构长廊改为了轻钢结构。

craftsman like carpenter or mason charges 300 *yuan* per day and an unskilled worker costs at least 160 *yuan* per day. Being on a limited budget, it's necessary to speed up the construction process as much as possible. That's why the material for the wall had to be switched from rammed earth to hollow bricks seamed with yellow mud, and the original wooden corridor had to be replaced by one made in steel which is much cheaper.

场地原状 I Original Site

上吴村蔬菜采摘设施用房

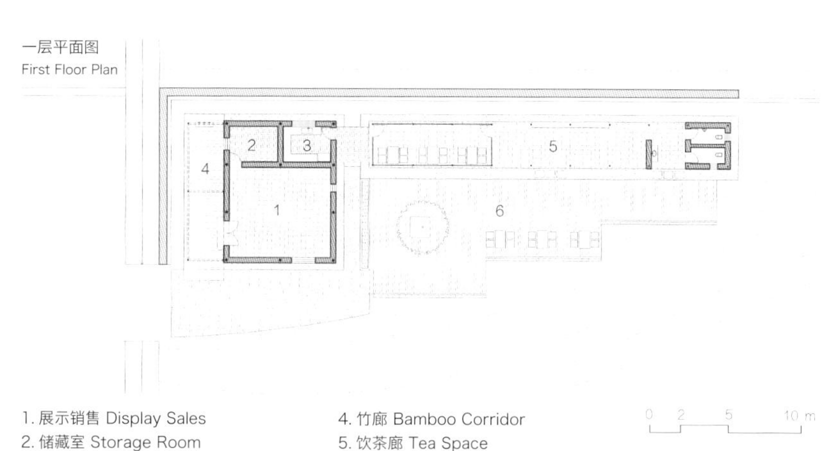

一层平面图
First Floor Plan

1. 展示销售 Display Sales
2. 储藏室 Storage Room
3. 茶水间 Tea Room
4. 竹廊 Bamboo Corridor
5. 饮茶廊 Tea Space
6. 观景平台 Terrace

建造过程 | Building Process

鄣吴镇公交站

Bus Station of Zhangwu Town

安吉县鄣吴镇 | Zhangwu Town, Anji County

建筑师：贺勇，张艳颖，陈耀
建筑面积：232 m²
结构形式：轻钢结构＋混凝土结构
建成时间：2016 年 2 月
Architects: He Yong, Zhang Yanying, Chen Yao
Gross Floor Area: 232 m²
Structure: Light Steel Structure + Concrete Structure
Date of Completion: February 2016

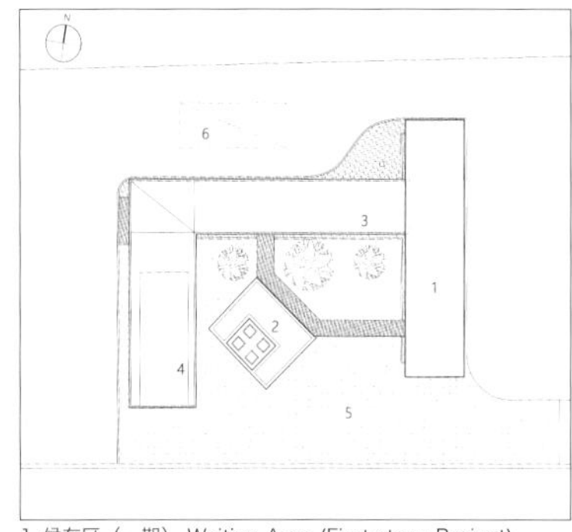

1. 候车区（一期） Waiting Area (First-stage Project)
2. 公共卫生间 （一期）Public Toilet (First-stage Project)
3. 候车区（二期） Waiting Area (Second-stage Project)
4. 值班休息室（二期） Duty Office (Second-stage Project)
5. 竹林 Bamboo Grove
6. 公交车停靠点 Bus Stop

总平面图 Site Plan

该项目分为两期完成：第一期是东侧轻钢结构的候车空间以及白色方盒子的公厕；第二期则是西北侧“L”形的清水混凝土顶棚的候车空间。

第一期的空间根据使用的功能不同呈现三种状态：开敞的等候空间（廊），半开敞的躲避风雨的玻璃盒子空间（亭），封闭的厕所空间（室）。等候空间部分的竹子是悬挂于顶部钢梁之上的，下部可自由活动，于是当一阵风吹过来的时候，候车人的耳边响起的不再是嘈杂的噪声，而是悦耳的竹风铃。厕所部分的面积很小，却依然想给人一点点惊喜：在那个外表干净的纯白小盒子里，当你推开最为私密的厕所蹲位隔间的小门时，粉红（女厕）、钴蓝（男厕）的强烈色彩扑面而来。当你蹲下时，抬头透过天窗可以看见蓝天与白云。

第二期的清水混凝土墙体和顶棚分别采用了松木以及席纹竹胶板的模板。这也是我们之前在郭吴村书画馆、玉华村厕所等多个项目中使用过的材料与建造方法，因为有了前面的施工经验，所以在这里取得了比较理想的效果，建筑表面呈现出了丰富的肌理。其形体上依然是一个简洁的开敞廊子，其顶棚在转角处微微翘起，以便在廊下的人们看到远处完整的山脉景观。

This project consisted of two stages. The first stage included the construction of the waiting area made of light steel on the eastern side of the station and a public toilet like a white box. The second stage involved an L-shaped waiting area with a ceiling of fair-faced concrete on the northwestern side.

The space in the first stage can be devided into three different spatial structures based on their different functions: the opening waiting platform, the semi-open shelter with glass walls, and the fully-enclosed toilet. There are bamboo poles hanging down from a steel crossbeam on the ceiling while their bottom ends are free from any attachment. The space of the toilet is quite limited but it's not without surprises. It looks like an ordinary whitewashed box, but when you open the little stool doors, the bright colors will flood into your eyes. Lift your head and you will be able to look through the skylight and see white clouds drifting around.

In the second stage, when building the concrete wall and the ceiling of the waiting area during the second stage, moulds made of pine wood and a kind of bamboo plywood with Xi Wen marking were used. This special material and building method have already been applied to previous projects multiple times, which is why it has created an ideal effect—the surface of the building presents a rich texture. Its general form remains a simple open veranda but some tiny changes to the details create a special effect. For instance, all the corner edges are tilted up which allows people to have a wider range of vision so that they can see the complete mountain landscape in distance.

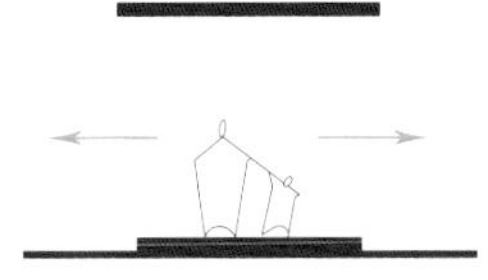

廊（公交站台）
沿水平方向全开敞，视线通透，公共性极强
Corridor (Waiting Area)
fully opens along the horizontal direction, with clear vision and strong publicity

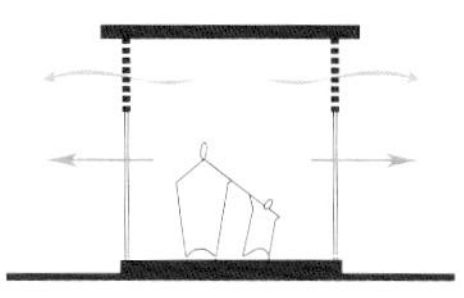

亭（室内等候区）
沿水平方向视线通透，顶部局部开敞通风，公共性强
Kiosk (Indoor Waiting Area)
has a clear vision on the horizontal direction with strong publicity, the top of which is open for ventilation.

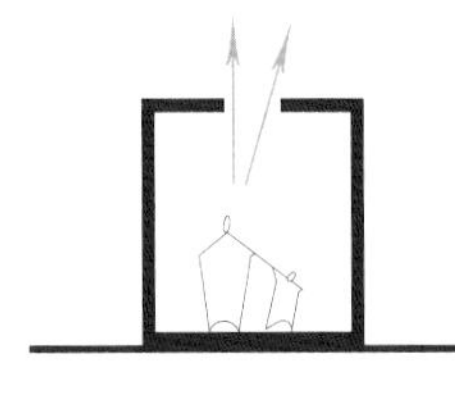

室（公共卫生间）
沿水平视线有阻隔，却向天空打开，私密性强
Room (Public Toilet)
cut off the horizontal line of sight, but open to the sky. Thus, the privacy is strong

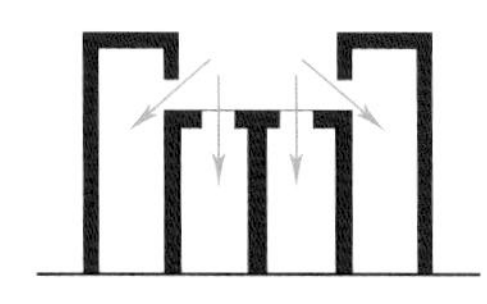

天光洒进封闭的卫生间内
Sky light
spills into the enclosed stalles

白色的卫生间盒子包着蹲位
The white box wraps up the toilet

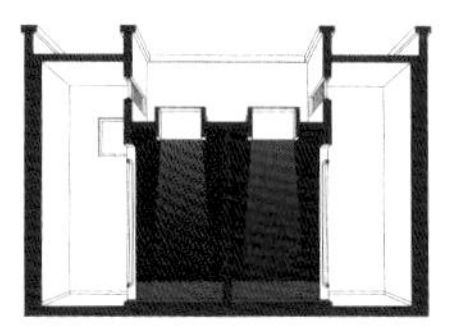

内部是彩色的蹲位，一束天光洒下
The inner toilet with a beam of sky light

郭吴镇公交站

J02
鄣吴一上堡
6:30-17:00
中国·鄣吴

轴测图
Axonometric Drawing

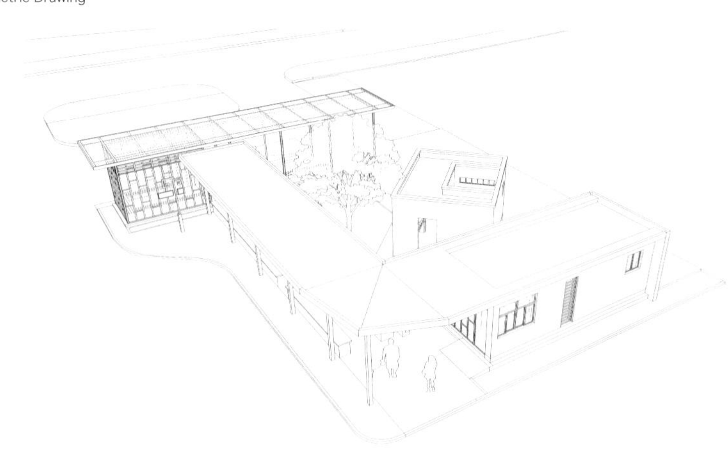

平面图
First Floor Plan

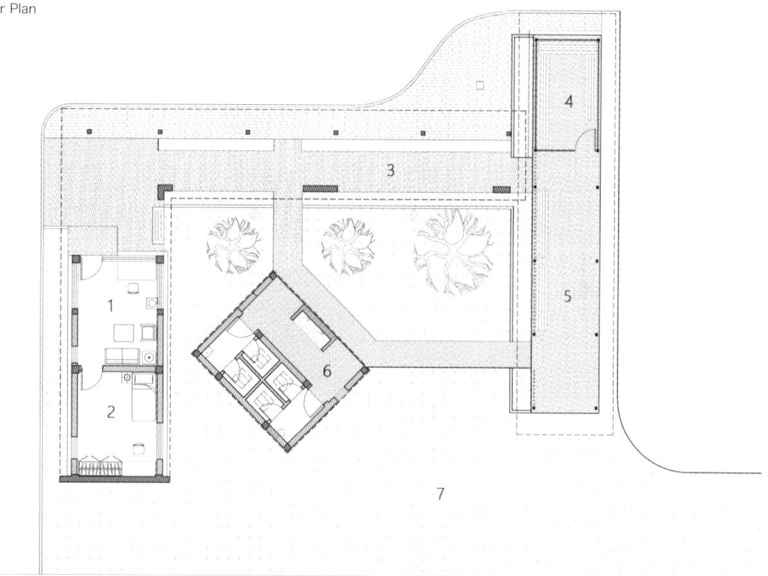

1. 值班室 Duty Office
2. 休息室 Rest Room
3. 站台等候区 Outdoor Waiting Area
4. 室内等候区 Indoor Waiting Area
5. 站台等候区 Outdoor Waiting Area
6. 公共卫生间 Public Toilet
7. 竹林 Bamboo Grove

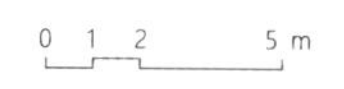

剖面 1-1

Section 1-1

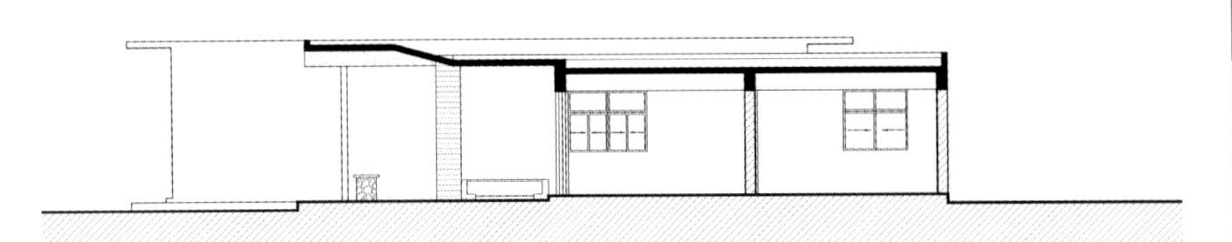

剖面 2-2

Section 2-2

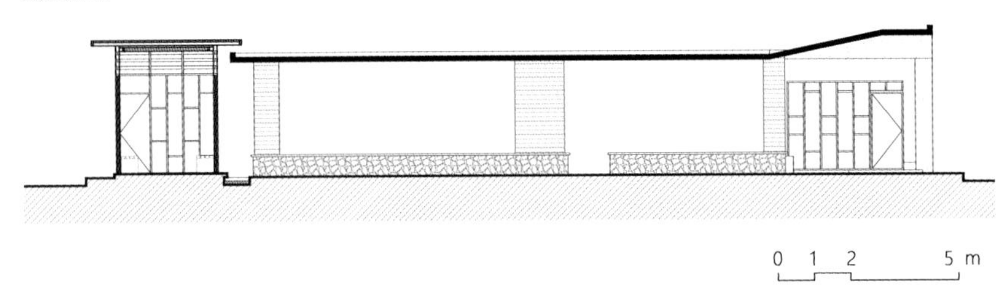

鄞吴镇公交站

郭吴镇公交中心

207

ARROW

ARROW

鄣吴镇卫生院

Zhangwu Town Hospital

安吉县鄣吴镇 I Zhangwu Town, Anji County

建筑师：贺勇，张艳颖，安吉县昌硕建筑设计院（施工图）
建筑面积：2 616 m^2
结构形式：混凝土框架
建成时间：2016 年 5 月

Architects: He Yong, Zhang Yanying, Anji Changshuo Architecture Design Institute
Gross Floor Area: 2,616 m^2
Structure: Concrete Frame Structure
Date of Completion: May 2016

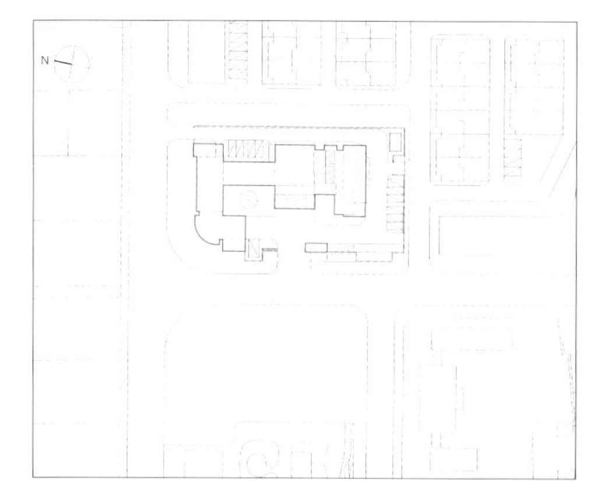

总平面图 Site Plan

高耸的大山以及优美的田园风光是人们到达鄣吴时所见的第一印象。在如此的环境中如何安放一座建筑，首要考虑的是建筑所呈现出来的尺度感，作为鄣吴镇目前所建的最大体量的建筑，它的出现将在很大程度上重新界定这里的建筑与环境之间的关系。为了延续周边建筑与场所之间既有的亲和关系，建筑师在此将建筑体块化整为零，再以一种紧凑的方式重新组合在一起，让其形体自然地呈现出聚落式的特征，从而以一种亲和的尺度融入山水田园与村落之间。

医院，可谓是功能性最强的一类建筑。尽管作为一家小型医院，该建筑设计过程依然遵循着严格的功能要求和相关规范。在极其有限的用地上，卫生院的主要功能模块——门诊、病房、急诊和预防保健相对独立布置，用廊道将它们连接起来，形成相对半集中式的功能布局。廊道是医院建筑不可或缺的部分，但往往形成单调的空间。在该建筑中，建筑师尝试对于通常的走道进行化解与利用，

My first impression for the town of Zhangwu is its magnificent mountain and breath-taking idyllic scenery. To insert a building into such an environment, the first question needed to be answered is what exactly the building has to present. As the biggest project Zhangwu has ever invested, this building will to a great extent re-define the relationship between its natural environment and the man-made architecture. In order to maintain the current harmony between the construction site and the surrounding buildings, the architect decided to divide the whole building into several pieces before re-connecting them and the nearby buildings into a more compact unit with a certain aggregational feature. This way, the new hospital would be able to integrate into the natural background of the mountain and the village.

The hospital is the most functional building. Though it's only a small-town hospital, the designing and building process still need to serve all the functions of a hospital and follow the related rules and regulations. That means that despite the extremely limited space, all of its main functional units, namely the outpatient service, the patient ward, the emergency room and the prevention and health section need to have

形成连廊、休息厅、内院环廊等多种高效、有趣的空间。围墙，是当下“中国式医院”的必需，但经过起承转合、虚实变化之后，使之与建筑的主体能够相得益彰，浑然一体。

受制于极其有限的投资和当地的建造水平，建筑的每一个部分基本都采用方正简洁的形体。该地的传统建筑在风貌上总体而言属于徽州民居，但是该建筑除了在尺度与色彩上与之相协调之外，形体上并未在建筑细部、符号上去刻意地表达或呼应，而是尽可能干净、简洁，旨在体现该建筑的功能需要与当下的建造方式和审美习惯。

their independent space and settings while connected with each other by corridors. This would require a half-centralized distribution of the functional units. The corridor is an indispensable component of a hospital and yet the form and space created by them tend to be quite monotonous. In this project, the architect attempted to give several sections of the corridor some extra functions by making them into a variety of highly efficient and appealing space such as a hallway, a rest parlor, or an inner court veranda. A bounding wall is an imperative component of the current "Chinese-style" hospital. With some twists and turns as well as interplay between the occupying and emptiness, the wall and the main building seems to be able to intermingle with as well as complement each other in harmony.

Due to the restriction of the limited budget and construction condition, almost every part of the building adopts a simple and square-shaped form. The local traditional architectural style is generally in line with that of "Huizhou" (Now Anhui Province) residence. However, unlike Huizhou-style residence, this building doesn't show any expressing or coordination in detailed decoration or symbols. Instead, it looks quite clean and simple aiming to express its functions and the current architecture style as well as the modern aesthetic features.

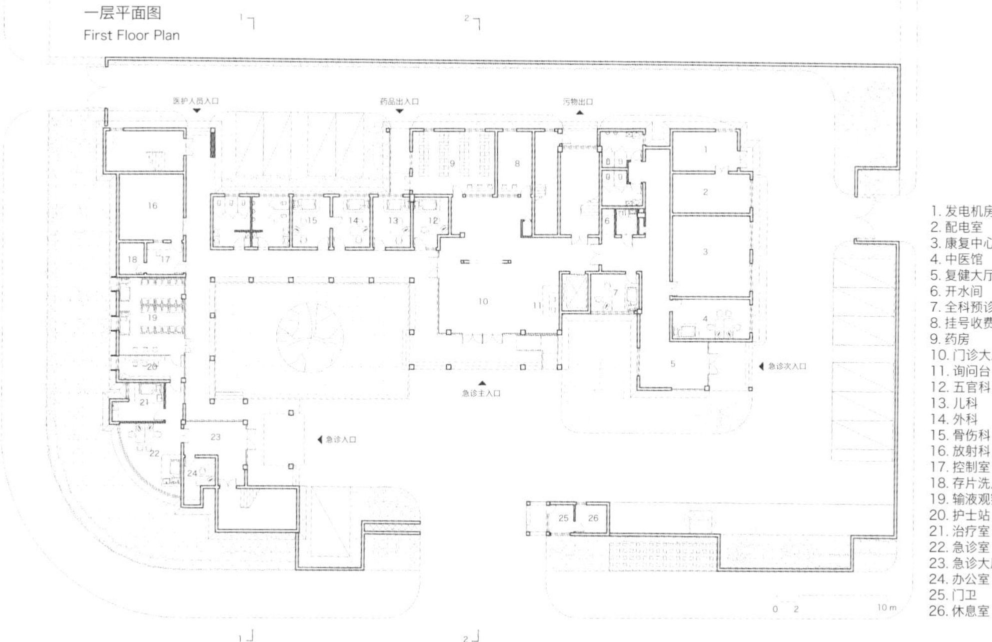

1.	发电机房	Generator Room
2.	配电室	Power Distribution Room
3.	康复中心	Rehabilitation Center
4.	中医馆	Chinese Medicine Center
5.	复健大厅	Rehabilitation Hall
6.	开水间	Water Room
7.	全科预诊	General Pre-diagnosis
8.	挂号收费	Registration
9.	药房	Pharmacy
10.	门诊大厅	Outpatient Hall
11.	询问台	Information Desk
12.	五官科	E.N.T.Department
13.	儿科	Pediatrics
14.	外科	Surgery
15.	骨伤科	Orthopedics
16.	放射科	X-ray Room
17.	控制室	Control Room
18.	存片洗片	Storage
19.	输液观察	Infusion Observation Room
20.	护士站	Nurse Station
21.	治疗室	Treatment Room
22.	急诊室	Emergency Department
23.	急诊大厅	Emergency Hall
24.	办公室	Office
25.	门卫	Guard
26.	休息室	Lounge

剖面 1-1

Section 1-1

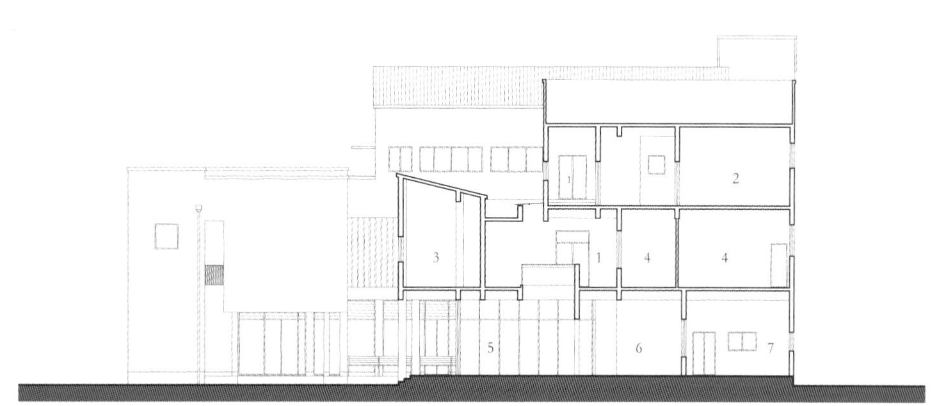

1. 走道 2. 产房兼手术室 3. 内科 4. 检验室 5. 门厅 6. 大厅 7. 药房

1. Corridor 2. Delivery Room and Operating Room 3. Internal Medicine 4. Test Room 5. Lobby 6. Hall 7. Pharmacy

剖面 2-2

Section 2-2

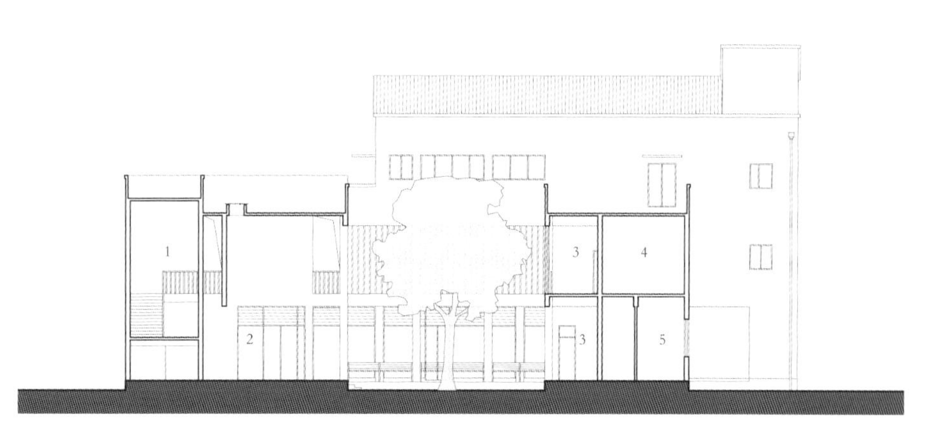

1. 楼梯间 2. 急诊大厅 3. 走道 4. 预防保健室 5. 女卫生间

1. Staircase 2. Emergency Hall 3. Corridor 4. Preventive Health Care Room 5. Women's Toilet

安全通道

安全通

景坞村村委会

Community Center of Jingwu Village

鄣吴镇景坞村 | Jingwu Village, Zhangwu Town

建筑师：贺勇，纪敏，浙江安吉县建筑设计研究院（施工图）
建筑面积：1 177 m^2
结构形式：混凝土框架
建成时间：2017 年 1 月

Architects: He Yong, Ji Min, Zhejiang Anji County Architecture Design and Research Institute
Gross Floor Area: 1,177 m^2
Structure: Concrete Frame Structure
Date of Completion: January 2017

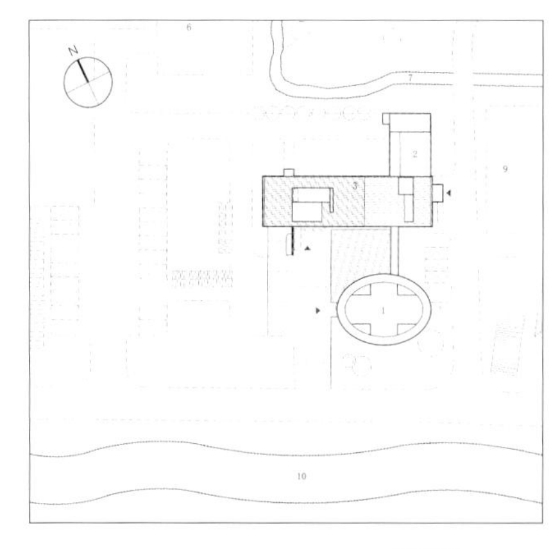

总平面图 Site Plan

该建筑用地的前面是一条小溪，背后是连绵起伏的群山，风景极好。初次见到这块场地，一个大大的“人”字坡屋顶就隐约浮现在脑海，最后建成的房子也的确如同一个依山傍水的大农居。名为村委会，其实也是社区的服务中心，包含了村委会办公、便民服务、医疗卫生站、农产品展示 / 旅游接待以及大会议室等多种功能。经过对场地条件以及功能组织进行分析，在建筑的总体布局中，将村委办公、社区服务这一主要功能单元用房（主楼）与旅游 / 展示用房相对独立布置，用廊道连接，彼此围合出一片水面和竹林景观。

主楼二楼北侧是外廊，临山，这里的外墙用灰砖镂空砌筑，形成大大小小的洞口，从这些洞口望出去，能看到不远处如画的田园风景。主楼三层利用“人”字坡中间的高大空间作为大会议室，两端则形成了两个屋顶露台。露台既是从楼梯间通往大会议室的必经之路，也为村民欣赏山间美景提供了新的平台。村委会建成后不久，周边陆陆续续修建了篮球场、室外舞台、文化展示长廊等文体休闲设施，目前这里已经是景坞村最有活力的公共场所了。

This construction site is fronted by a creek and backed by the rolling mountains, offering a magnificent scenery. The first time I visited this place, an image of a giant Y-shaped roof immediately hopped into my head, which was exactly what's been constructed in the end, as a part of a big farm-style house surrounded by rivers and mountains. Though named "the village council", this house also serves as a community service center with multiple functions and facilities for medical service, farm products exhibition, tourist reception, and conference room. After a thorough analysis of the physical condition and the functional structure of the site, we came up with a general idea about the layout. A corridor will be built to connect them and to create an enclosed garden scenery with water pools and bamboo patches.

The northern side of the main building lies a verandah facing the mountain. This section of the bounding wall adopts a "hollowed out" design using grey bricks. Through numerous holes on the wall, we could have a view of the picturesque landscape in the near distance. The third floor is right below the big Y-shaped roof. With such an unusually high ceiling, the space for this floor is much bigger than other rooms and therefore is perfect for a conference room. There are two rooftop terraces at the two ends of the room, one for each end, which are part of the only route leading from the staircase to the conference room and provide new platforms for the villagers to enjoy the mountain scenery. Soon after the construction was completed, more cultural and entertaining facilities popped up in the surrounding area one by one, such as a basketball court, an outdoor performance stage, and an art exhibition corridor. Now, this area has become the hottest spot for public activities.

场地原状 | Original Site

设计构思 | Design Concept

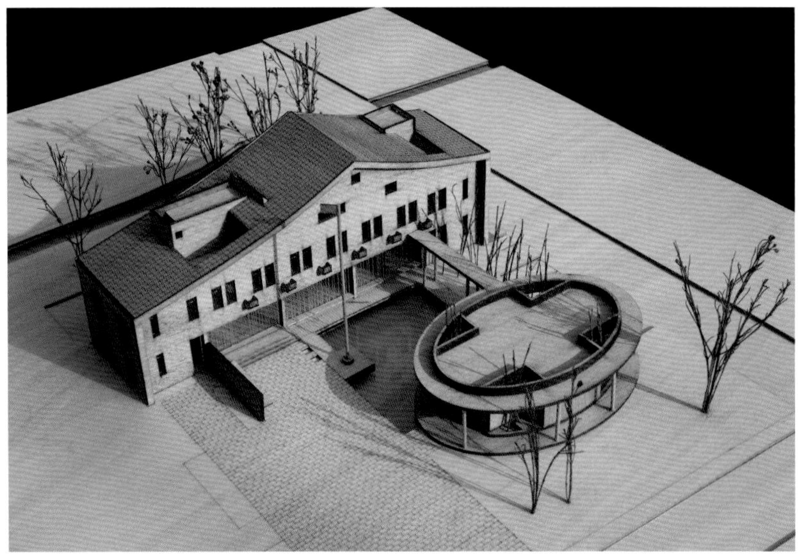

模型 | Model

积极培育和践行社会主义核心价值观
郭吴镇景坞村经济合作社
郭吴镇景坞村村民委员会
中共郭吴镇景坞村支部委员会
郭吴镇景坞村村务监督委员会

景坞村村委会

一层平面图

First Floor Plan

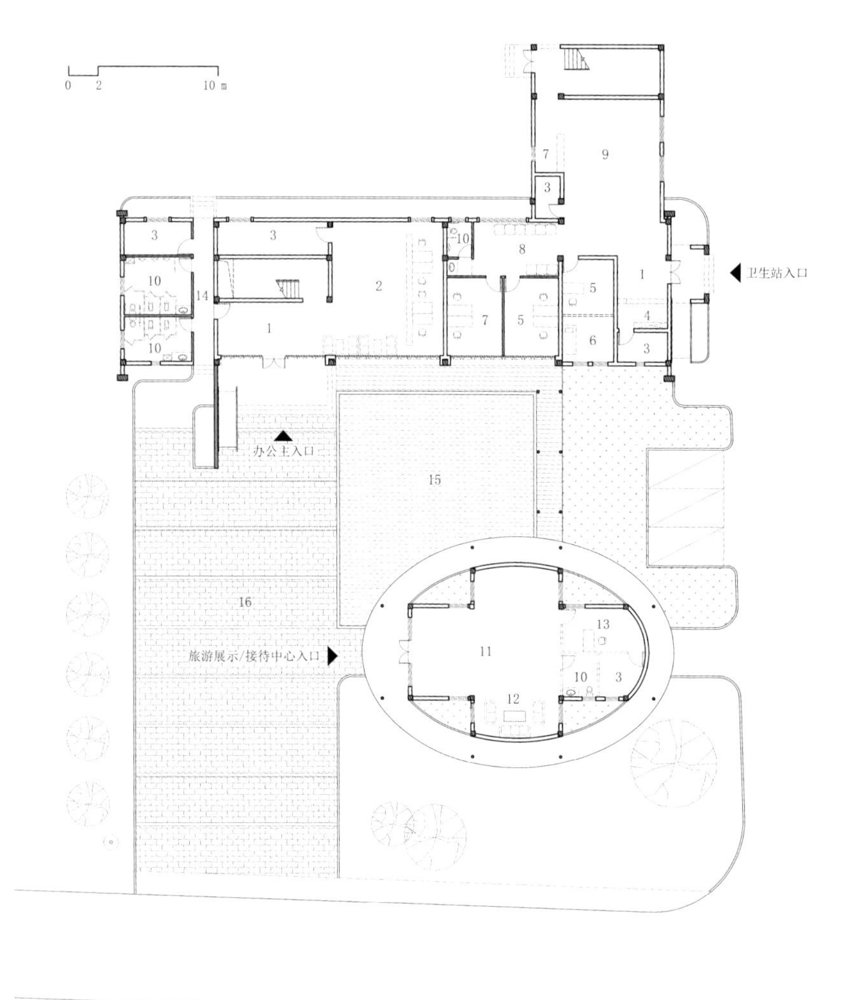

1. 门厅	Hall	2. 办事大厅	Service Hall	3. 储藏室	Storage Room	4. 药房	Pharmacy
5. 医生办公室	Doctor's Office	6. 检查室	Check Room	7. 护士办公室	Nurse's Office	8. 等候区	Waiting Area
9. 输液区	Infusion Area	10. 卫生间	Toilet	11. 展示大厅	Exhibition Hall	12. 接待	Reception
13. 办公室	Office	14. 通道	Passage	15. 水面	Water	16. 休闲广场	Square

景坞村村委会

剖面
Section

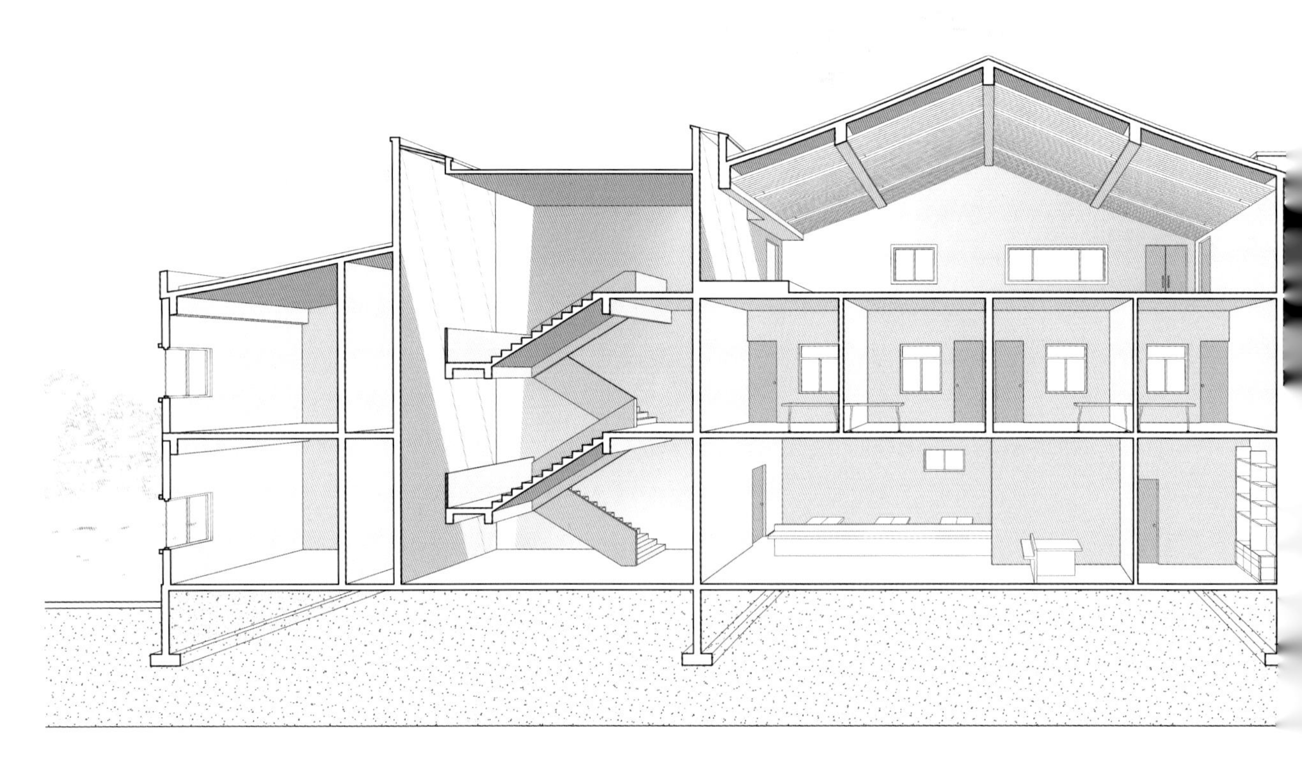

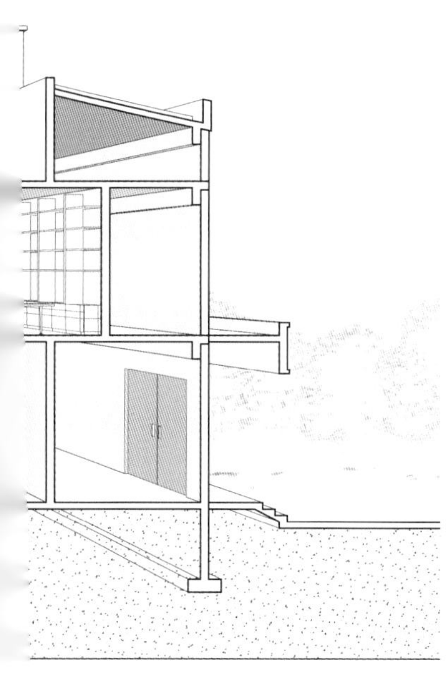

服务中心

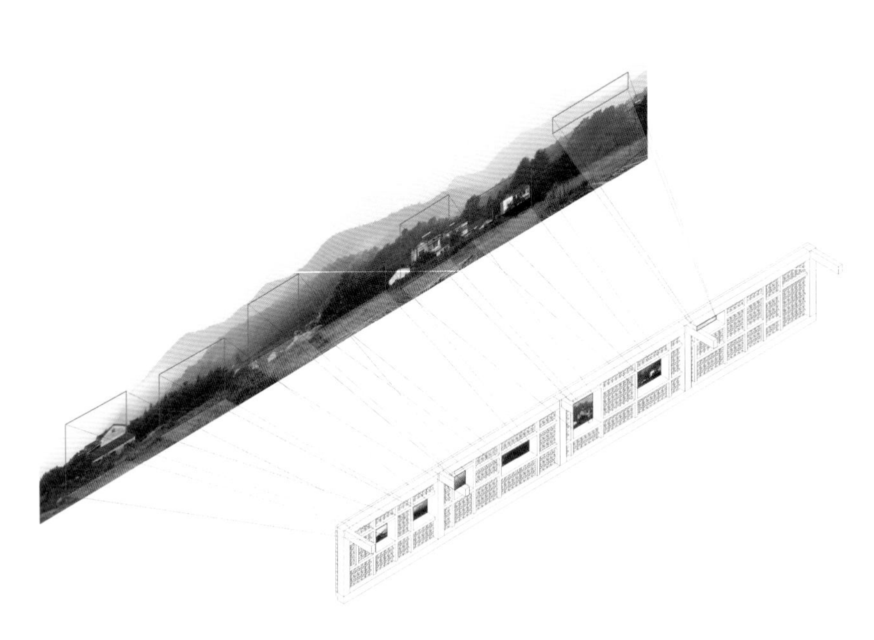

鄣吴镇景坞村土特产展示中心

年味进家
欢乐进家庭 文明

玉华村村委会

Community Center of Yuhua Village

鄣吴镇玉华村 I Yuhua Village, Zhangwu Town

建筑师：贺勇，孙姣姣，浙江安吉县建筑设计研究院（施工图）
建筑面积： 756 m²
结构形式： 混凝土框架结构
建成时间： 2017 年 3 月
Architects: He Yong, Sun Jiaojiao, Zhejiang Anji County Architecture Design and Research Institute
Gross Floor Area: 756 m²
Structure: Concrete Frame Structure
Date of Completion: March 2017

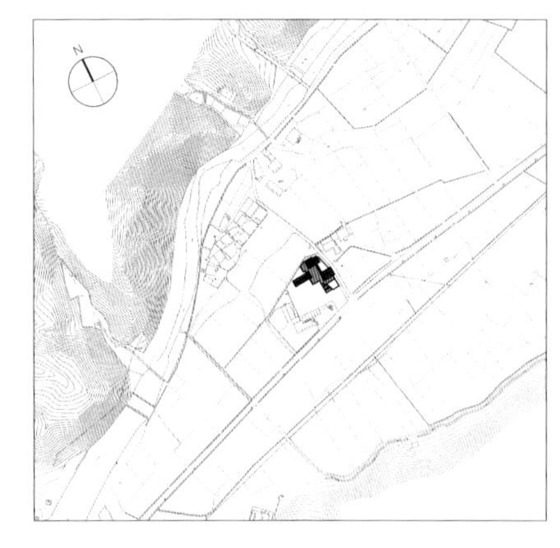

总平面图 Site Plan

背靠高耸的山脉，门口一片种植不久的樱桃园，透过浓密的树枝，隐隐约约可见一组农居，这便是玉华村的所在。该村位于浙江省安吉县鄣吴镇，毗邻国家级历史文化名村鄣吴村，地处金华山和玉华山之中，青山环抱、绿竹满山，全村人口约850人。其经济来源主要是茶叶、笋以及休闲农家乐等。玉华村村委会位于玉华村中心，背后是一组新近完工的农居，门口有一条穿越该村中心的乡道，这也是该村与外界联系的主要通道。玉华村村委会原本就在这个场地之上，20多年前建的2层小楼过于狭小、破败，所以需要拆了重建。一栋一层的条形长屋是原来的活动室，质量尚可，于是保留了下来。除此之外，一个篮球场、一座亭子、四棵大树，便是场地上的全部要素。

村里经济条件有限，房子不能随意地拆，更不能任性地建，所以建筑的主体基本是原拆原建，但毕竟面积增加了不少，所以如何处理新建筑与保留的构筑物的关系成了设计中的难点。篮球场是村委会原来主要的户外空间，与

Backed by the magnificent mountains and fronted by an orchard of newly-planted cherry trees, a row of village houses is barely visible through thick tree branches. This is where Yuhua village is located, belonging to the Zhangwu Town of Anji County in Zhejiang Province, adjacent to the village of Zhangwu, between two hills named Jinhua and Yuhua and covered by verdant bamboo trees. It has a population of about 850 whose main income source is the business of tea & bamboo shoots trade as well as its agri-tourism. The new community center of Yuhua village is located right at the center of the whole village, in front of a group of newly-built village houses, and next to the main road running through the village center. The old community center used to be at the same location. However, being built more than 20 years ago, the old two-storey building has dilapidated a lot and there was simply not enough space. Therefore, remodeling and expansion are needed. The previous activity room on the first floor is the only part that has been preserved. Apart from that, there are other elements available for the project: a basketball court, a pavilion as well as four big trees in the courtyard.

原来的建筑在大体上呈“L”形咬合在一起，所以在新的布局中，维持了这样的格局，从而实现了新、老建筑以及与篮球场空间关系之上的延续。

场地上的几株香樟、玉兰的树龄都已经是20年以上，树干高大，枝叶繁茂，是场地上最具生命力的景观，它们见证了该村在过去几十年中发生的变迁，没有理由被砍伐甚至移植。于是，如何保留几棵大树，使其能与新的建筑融为一体，成了设计中的另一个关键所在。设计之初，我们对于大树的位置、树干大小、树冠所覆盖的半径等都进行了仔细的测绘，在设计之中，遵循着每棵树的特征，通过恰当的建筑语汇，将其整合到围合院子的边界、路径的中心、入口的对景之上。这样一方面既保留了大树，也让其与新建筑融合在一起。

建成后的村委会建筑因尊重了大树和周边的农居，其自身一个完整的立面都难以呈现出来。可是，在这个退让、消融之中，房子却被牢牢地锚固在了场地之上，呈现出一般新建筑少有的妥帖与安定，自然成为该村景观，乃至日常生活中的一员。对于村里的一位“新来者”，或许这是遵循场地原有脉络的最好结果。

中国当下的乡村，村主任经村民直选产生，其管理上的民主程度可谓达到历史最高，但是由于长期以家庭为单位的生产经营方式以及很多村民进城经商、务工的现实状况，村民们普遍对村集体事务表现淡漠也是不争的事实。作为一个人口不足1 000人的小型村庄，村委会在满足了日常的便民服务、村务管理、文化礼堂等这些基本的功能之后，其实也并不需要更多的房子。建筑师思考的重点是如何通过这个建筑，在此创造更有吸引力的公共空间，激发更多的公共生活。尽管村里的公共与集体生活并不是很活跃，但是打篮球、地方戏曲的表演、书画社的交流等这些活动还是有一些的，城里流行的广场舞在此也颇受村民们喜欢，所以如何满足此类活动需要，激发更多此类的活动，成了设计的重要目标。

作为村里的公共服务与公共生活的中心，开放、通达是其必要的空间性格，其设计策略具体体现在以下三个方面：首先，场地之上以开放的空间为主、建筑为辅，即篮球场成为真正的主要空间，因为这里承载了村民最多的日常活动；其次，在建筑之中，有多个对外开放的院子，院子与房子之间有着宽敞的连廊空间，它们共同构成层次丰富的多种空间状态，成为联系室内外活动、满足多种功能

Due to the limited budget, it is unlikely that we could just tear down or construct buildings at will, so one of the biggest challenges for us is how to deal with the relationship between buildings that are newly-built and those preserved ones. The basketball court is one of the major outdoor space at the old community center which forms an L-shape with the old building. In the new design, we have kept the layout in order to maintain the spatial relationship between both old and new buildings as well as the basketball court.

There are several camphor trees and magnolia trees at the community center, which are more than 20 years old. Blessed with sky-reaching height and a luxuriant canopy of flourishing branches and leaves, these trees have witnessed all the ups and downs of this place so any attempt of cutting down or even moving them to somewhere else is simply out of the question, which means how to integrate the new buildings with these trees has become a key issue to this project. At the very beginning, we have made a thorough examination and measurement as to the position, the size of the tree trunk as well as the diameters of the tree crowns, based on which, we made new spatial arrangement to the courtyard in such a way that these trees are now respectively located at the edge of the courtyard, the middle point of the pathway and the entrance of the community center. In this way we managed to fit the trees and the new buildings together.

Due to the necessity of keeping the trees and the surrounding village houses undisturbed, the newly-built community center failed to show its own facade in its entirety. However, during the process of retreating and melting into the environment, it has been firmly anchored to the ground, presenting its fittingness and stability that are rare for newcomers.

Currently, in Chinese rural villages, all village committee directors are directly elected by all the villagers and the daily operation of public affairs are highly democratic. However, due to the fact that most of the daily production activities are family-oriented and that more people have left their home to seek employment in cities, villagers tend to be much less concerned about the village collective affairs. Therefore, as a small village with a population of less than 1,000 people, Yuhua village committee only need space for the most basic functions, such as a public service center, a village affairs management office and an auditorium for cultural events. What the architect needs to focus on is how to create a more appealing public space in order to promote more public activities. In actual fact, there has been some popular pubic activities like basketball matches, local traditional opera shows, as well as the regular communication and discussions in the calligraphy & painting club. Besides, the square dancing which is quite prevailing in cities these days is also very popular among villagers. Therefore, an important goal of this project is to find ways to provide enough space for activities of such kinds.

As the center of public services and public activities in the

的重要场所；最后，一条从一层屋面上可穿越该村委会的路径，也为村民的到达或离开提供了另一种可能。

总之，一个篮球场（兼作广场）、几个院子、用于展示的外廊、大台阶之上的平台，构成了一个富有趣味的乡村社区中心的空间，提供了满足多种户外活动的场所。当村民来此办事、活动，透过围墙上的孔洞、室内的大玻璃窗，或者站在一层屋顶的平台，不经意间，他们会发现自家村子中许多曾经“熟视无睹”的美景，从而感到一丝惊喜。这在很大程度上也成为许多村民在此驻足的理由。值得一提的是，戏台背景墙上的巨幅梅花图是几位村民合力完成的，从构思到打底塑形，再到着色深化，经历了近一个星期的紧张工作；门口保留亭子的屋面的翻修也是由村里的工匠自己完成。这种村民参与的建造方式对于建筑公共场所氛围的形成也起到了积极作用。

当下在乡村里建房子，很容易陷入使用砖瓦、石头、木材这样的套路，其实，站在村民的角度，经济、实用才

village, its most fundamental spatial characteristics would have to be openness and connection and so the designing strategy is mainly reflected in the following three aspects. First, open space such as the basketball court is supposed to occupy the center stage while buildings are only in complementary roles. Second, the buildings, a number of courtyards and a spacious outdoor corridor connecting them will form a multiple-layered spatial arrangement which serves as a connection between indoor and outdoor activities as well as satisfying the demands for multiple functions. Third, there's a route along which people can walk across the whole community center, providing an alternative way for villagers to reach and leave this place.

All in all, a basketball court which also functions as the village square, several courtyards, an open corridor and a platform high above the ground have formed an intriguing space for the village community center, offering an appropriate place for outdoor activities of various types. Every time when people come here for business or leisure, they would be pleasantly surprised by the fact that their village which they used to find nothing out of ordinary look so refreshing and delightful when seeing through the holes on the bounding wall, the vast glass window or from the terrace on the rooftop of

是其最自然也最基本的建房态度，所以选用当下最常用的材料与建造方式是顺理成章的事情。玉华村村委会在形体与建造方式上，与其说是一个公共建筑，不如说更像一组大一点的农居，框架结构、填充砖墙、白墙灰瓦是其基本的建造方式与材料构成。不过，建筑总是要有点新意的，所以在院子的围墙上，外侧选用了竹子模板的清水混凝土，使其颜色、肌理和质感与建筑主体的白墙有所对比。选用竹子模板，一方面是因为该地盛产竹子，所以希望竹子的元素在建筑上有所体现，另一方面则是因为这种方法我们在该地其他项目中已经有所使用，而且达到了很好的效果。这次的竹子模板做法依旧与以前一样，即将直径8～10厘米的毛竹一分为二，然后钉在木工板之上。不过，源于上个项目中竹模板之间在竖向上接头效果不佳的教训，这次为了避免在垂直方向上有接头的痕迹，部分竹子模板长达近4米，这给模板的制作以及混凝土的浇筑都带来了很大的困难，因为如此长度的竹子，两端的直径已经有明显差异，

the community center. This, to a large extent, has also become the reason for many villagers to stay here. What's also worth mentioning is the huge plum blossom painting hanging on the outdoor performing stage serving as the background. This particular painting is the result of the cooperated work by several local villagers who have taken a whole week of hard work starting from scratch till its finish. The renovation of the pavilion at the entrance of the community center has also been done by local builders. It's believed that the participation of the local villagers in the construction process is instrumental in the creating a public atmosphere.

In rural villages, when building a house, the most commonly used materials are nothing more than bricks, tiles, stones and wood. Obviously, this choice is based on a principle of practicality and cost-effectiveness which is the most natural and basic attitude among the villagers. Yuhua village committee center is more like a group of slightly bigger farmhouses than a pubic building in terms of its structure and its construction style as well as its building materials. However, a bit originality is also necessary and that's why the road-facing side of the bounding wall is made by casting fair-faced concrete into bamboo moulding boards which creates a clear contrast

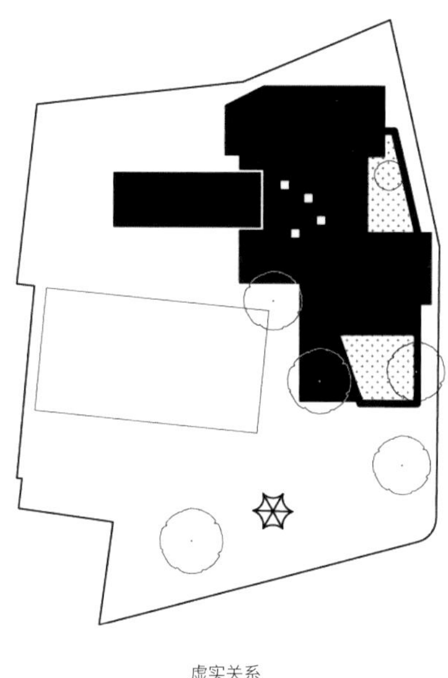

虚实关系
Soild vs Void

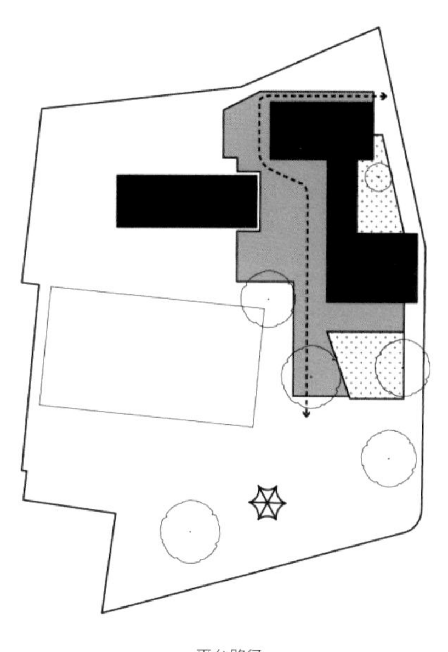

平台路径
Passage Way on the Terrace

所以需要精心的调整，使其一致，另外，一次浇筑如此高度的混凝土，也给其捣筑带来了很大的困难，需要工人在施工过程中更加的谨慎与耐心。由于商品混凝土加入了某些添加剂，其凝固后表面总是呈现出灰黄的效果，不甚理想，于是在本项目中我们选择了现场自拌混凝土，这也在一定程度上增加了工人劳作的强度。总体而言，完工后的清水混凝土墙体效果还是非常理想的，其沉稳的质感以及在阳光下呈现出的动人肌理，让建筑有了更多当下的时代气息。

经过了断断续续近一年时间的施工，玉华村村委会在2017年3月正式完工并投入使用。在这个建筑里，文脉、公共性不是概念，更不是口号，而是村民日常生活需求在场所与空间上的真实应对。其设计在合理之中暗藏着意外，熟悉之中显出一点陌生。另外，在乡村的在地建造中，在与村民的交流过程里，不提“文化”“文脉”这样一些大的概念，转向如何处理新老建筑的衔接，以及如何保留场地上的大树等这些具体入微的事情，更容易获得大家的理解与支持。在讨论、解决这些看似微小的事情的过程中，建筑自身也获得了独特性。玉华村村委会没有一般通常村委会的那种气派或强烈的标志性。从远处看过来，与周边农居几乎完全融合在一起，你几乎都分辨不清楚哪栋房子是村委会，但只要看到那几棵大树，就会轻易地找到它，因为那些大树已经存在在那里数十年。这种源自村民心底的记忆与认同，或许才是将人们从自家小院吸引到此地的力量吧！

with the whitewashed wall of the building itself in color and texture. The choice of bamboo mould is partly due to the fact that bamboo is a local specialty which should be an element reflected on the local architecture, and partly because this particular method has been successfully used in other projects. This time the mould have been made in the same way as before. We started by splitting the unprocessed bamboo tubes with an average diameter of 8 - 10 cm into two curved bars before nailing them to a wooden board. The only difference is that the bamboo tubes we have chosen this time have an average height of nearly 4 meters in order to prevent the use of joints between two bamboo bars which don't look pleasant. However, it also means difficulty in the making of the moulding board as well as in the concrete-casting. Since there's bound to be a big discrepancy between the widths of the top end and the bottom end of the unusually long bamboo tubes, we need to find a way to make up for the difference to make the whole board look neat and smooth. Due to the unusual height, the casting process became especially challenging. Workers had to be extremely careful and patient. Another problem concerns the concrete. Since those store-bought cement powder has some kind of additive in them, so after solidification, the surface will be in a grayish yellow color which is not what we want. Therefore, we decided to mix our own concrete right on the construction site which unfortunately made the builders' job even harder. However, it all worked out fine. The finished wall displays a solemn quality as well as a delightful texture especially under the sunshine which brings to the building a certain contemporary spirit.

After nearly a whole year of intermittent construction, Yuhua village community center was finally completed and put into use in March, 2017. In this project, "cultural heritage" and "publicity" are not just concepts or slogans, they are practical solutions to villagers' daily demands for space. There is unexpectedness hidden in rationality, and strangeness under the veil of familiarity. Another lesson that I have learned during my communication with the villagers is that refraining from talking about the abstract concepts such as "culture" or "legacy" but focusing on dealing with concrete problems like preserving the old trees or connecting the old and new buildings is much easier to gain understanding and support from the villagers. Looking from afar, Yuhua committee center does not particularly stand out with some kind of majestic signature.Instead, it looks just like one of those ordinary village houses. You cannot even tell them apart. However, once you saw those giant trees which have been there for decades, it would be extremely easy for you to find it. I believe it is this type of collective memory and recognition buried deep down in every villager's heart that has been a special force pulling people from their own houses to this spot.

华村
文化礼堂

玉华村村委会

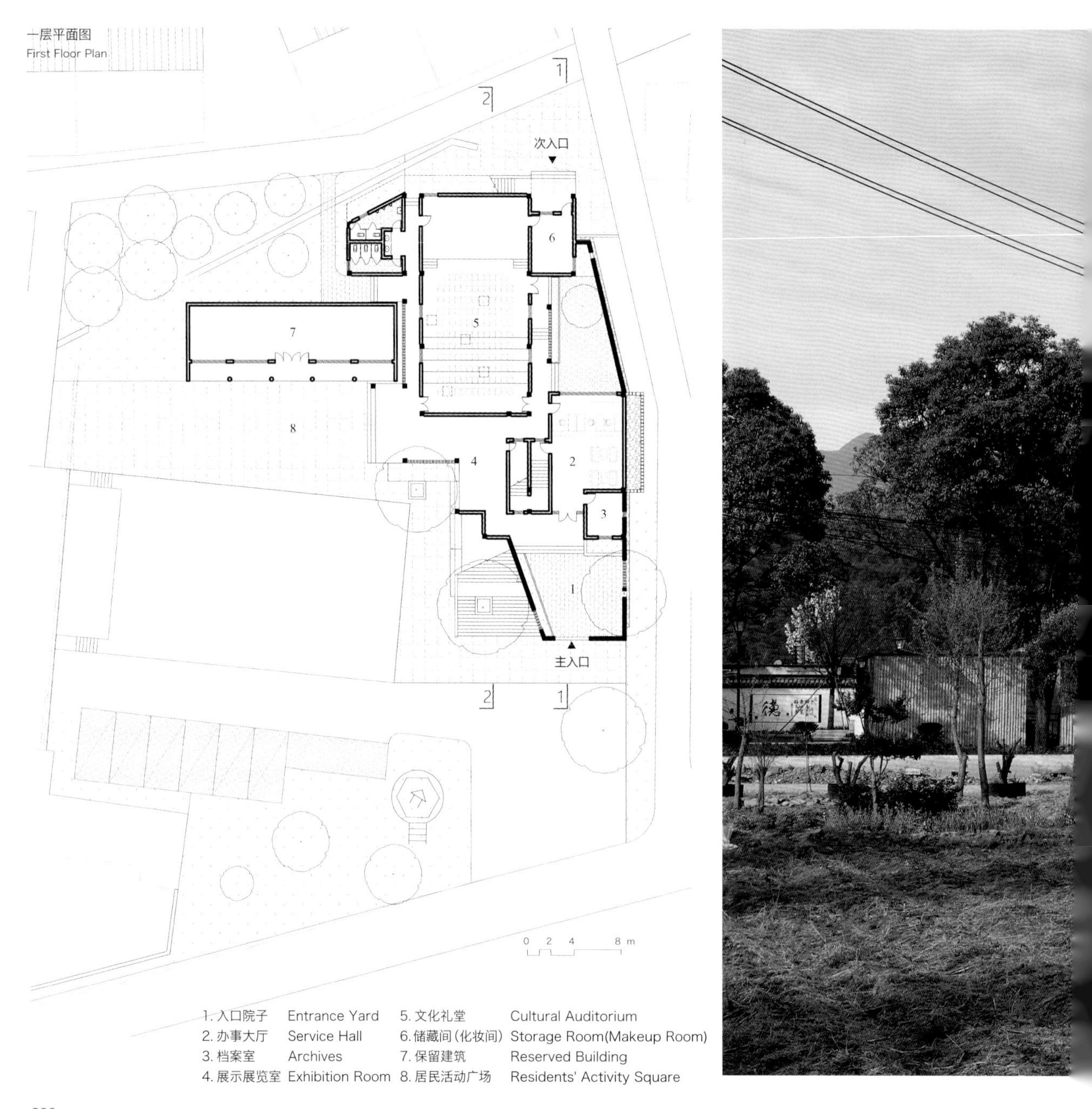

1. 入口院子 Entrance Yard
2. 办事大厅 Service Hall
3. 档案室 Archives
4. 展示展览室 Exhibition Room
5. 文化礼堂 Cultural Auditorium
6. 储藏间(化妆间) Storage Room(Makeup Room)
7. 保留建筑 Reserved Building
8. 居民活动广场 Residents' Activity Square

玉华村村委会

二层平面图
Second Floor Plan

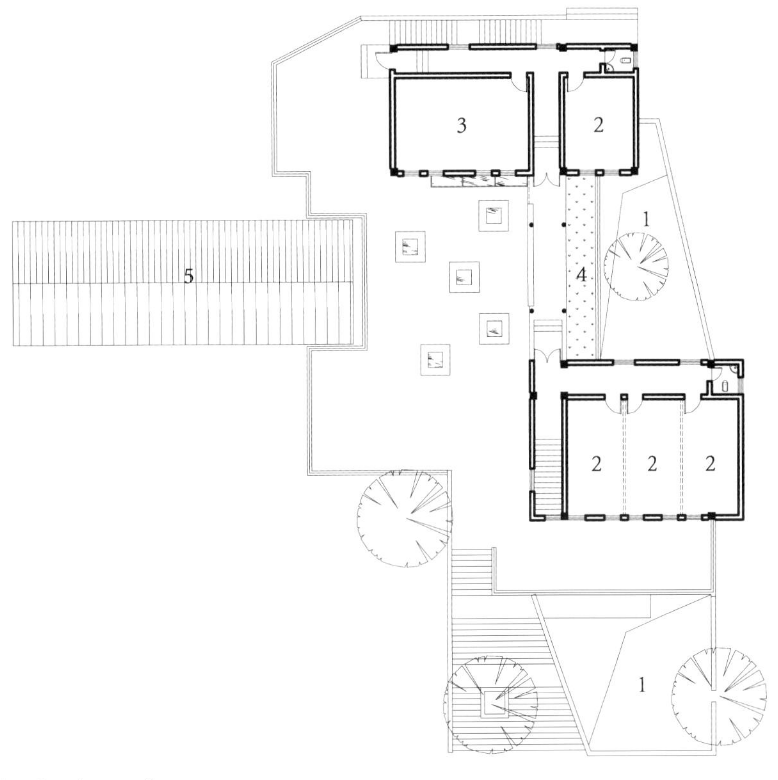

1. 院子上空　Over the Yard
2. 办公室　Office
3. 会议室　Meeting Room
4. 屋顶绿化　Roof Greening
5. 保留建筑屋顶　The Roof of Preserved Building

东立面图
East Elevation

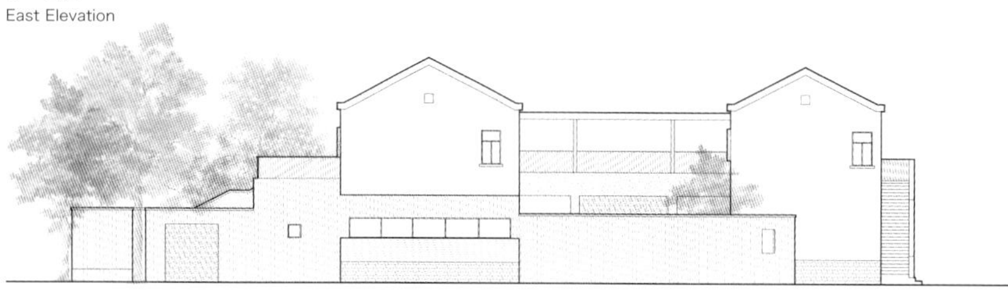

西立面图
West Elevation

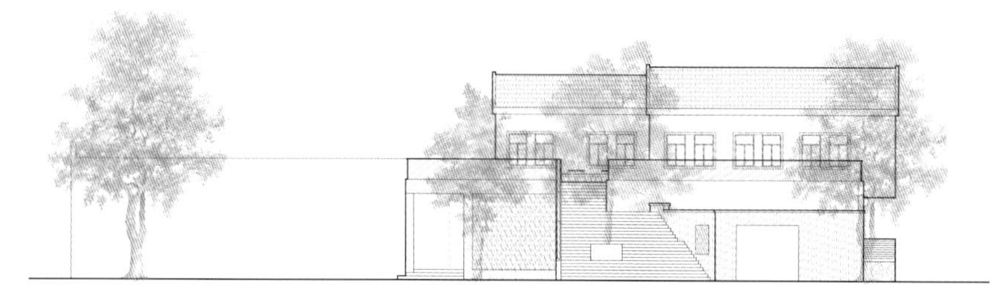

剖面 1-1
Section 1-1

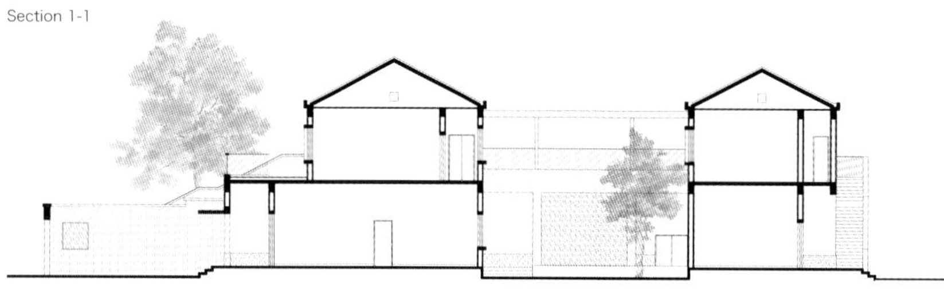

剖面 2-2
Section 2-2

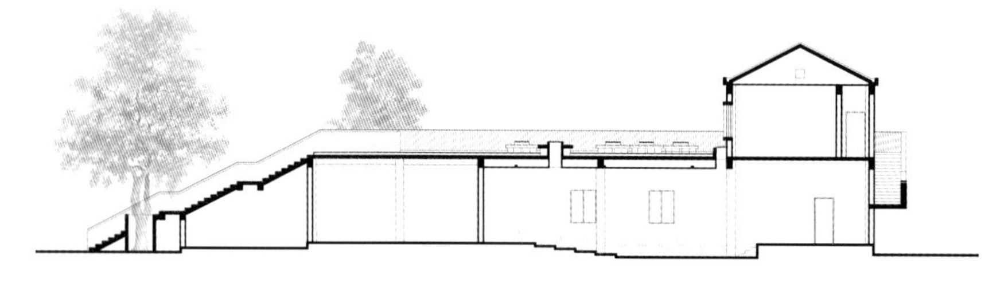

玉华村村委会

FLOWERS

玉华人家
微笑来敲门
睦邻情谊深
项目建设
文体活动

玉华村文化礼堂

玉华村庆祝重阳节特举办“夕阳美，邻里情”活动

玉华村竹酒设施用房

Bamboo-wine Brewery Building of Yuhua Village

郜吴镇玉华村 | Yuhua Village, Zhangwu Town

建筑师： 贺勇，戚骁锋，王凯伦，老陈
建筑面积： 274 m^2
结构形式： 钢结构 + 竹木结构
建成时间： 2017 年 9 月
Architects: He Yong, Qi Xiaofeng, Wang Kailun, Mr. Chen
Gross Floor Area: 274 m^2
Structure: Steel Structure + Bamboo & Timber Structure
Date of Completion: September 2017

总平面图 Site Plan

“小微”是乡村的一种基本特征，其不仅表现在建筑的尺度与规模上，也表现为空间与景观演变的速度，正是这种“小微”的状态，让乡村在建筑与景观风貌上呈现出了不同于城市的特征，显示了其独特性。如果把建筑作为一项“微设施”，其背后则往往对应着“微企业”“微整合”“微更新”，以及建筑师引导下的设计与建造方式。我们相信，通过这些微设施，乡村的产业得以不断提升，乡村的空间不断优化，乡村社会持续地改良。在这一理念与背景之下，玉华村竹酒设施用房可以作为一个代表性的案例。

两年前，老陈夫妇在浙江安吉县鄣吴镇玉华村租下了一片竹林，作为他们生产竹筒酒的基地。竹筒酒，就是将高度的纯粮酒在春天用针筒注入新长出的竹子里，一年后再将竹子砍下，然后敲碎竹节，倒出即可饮用。我品尝过几次，感觉不错，其酒精度数不高，另外酒中有股竹子的清香，味道很有些特别。

A basic feature of rural villages is "small"—the micro-scale of architecture, the small steps of changing in space and landscape. It is the very characteristic that distinguishes rural villages from cities. If we consider architecture as some sort of "micro-facilities", there will be corresponding to "micro-companies", "micro-integration" , "micro-regeneration" as well as a certain method of designing and constructing. We believe that these small facilities will be able to constantly boost the development of rural industries as well as its spatial arrangement. The bamboo-wine brewery building of Yuhua village is a case in point.

Two years ago, Mr. and Mrs. Chen rented a bamboo patch located at Yuhua village, Zhangwu Town, Anji County of Zhejiang Province, as the base for their bamboo-wine brewery. To make bamboo-wine, you need to inject pure rice wine with high alcohol content into the newly grown bamboo tube using a special syringe in spring. The next spring, the bamboo trees filled with wine will be cut down before the wine is poured for drink. I myself have had a taste of it once. To me, it tastes mild with a special flavor and a kind of bamboo-like aroma.

Being a wine-processing base, there's a need of buildings

作为竹酒生产的基地，需要一点临时设施用房，用以竹酒的加工、制作，以及展示性体验等功能。我们有幸参与了这个项目的设计。设计之初，老陈带着我们在那片竹林里考察，阳光穿过茂密的枝叶，在坡地上画出一道道光亮的弧线。竹林旁边有一大片刚刚种下的樱桃树，还有远处的山脉，在阳光下显得格外纯净，一副世外桃源的样子。老陈陪着我们一边走，一边描绘着他的设想，关于房子的布局、功能他都早已经成竹在胸。其实这块场地的特征是非常明显的，所以我们很快达成了共识，那就是在满足生产功能的前提下，需要让房子很好地融入场地，特别是要充分地利用北面那一片如画的田园风光。当然还有非常重要的一点，建筑的造价要尽可能低，因为老陈的竹酒事业才刚刚起步。

走出那片竹林，建筑的方案在我脑海里就已经基本形成：两个单坡屋面的长方形盒子顺着等高线布置，面向樱桃园的那片墙面是透明的玻璃，再加上一个出挑的外廊，

to house the brewery facilities as well as for exhibition. We have been fortunate enough to be part of this designing project. First, the owner, Mr. Chen gave us a brief tour around the bamboo patch which looks like some kind of utopian world with all of its natural splendor and beauty. We walked and talked about our ideas on the basic layout and functions of the new building. In actual fact, the geological feature of the building site is quite distinctive, based on which a mutual understanding was soon made between us. The basic idea is that on top of satisfying its production functions, we are aiming to integrate the building into its immediate environment seamlessly without bringing any disturbance to the picturesque natural scenery. Obviously, one more important principle is that the whole project has to be cost-efficient since Chen's business is still on its launching stage which means a limited budget.

The whole design has presented itself in my head as I was walking out of the bamboo patch. The basic function space will include two rectangular-shaped houses with single slope roof, a glass wall facing the cherry orchard and finally a verandah outside the house. Based on the undulation of the ground level, there will be several outdoor platforms and long stairs or a ramp way to connect the two small houses. After the first draft

场地原状 | Original Site

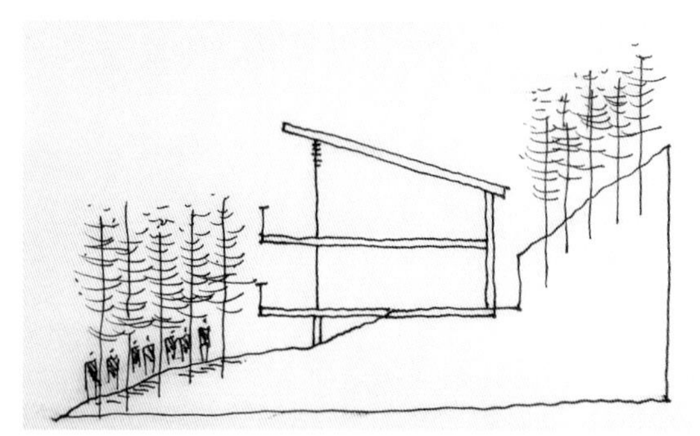

设计构思 | Design Concept

设计方案 | Design

模型 | Model

就构成了基本的功能空间。再结合场地的标高变化，用几个户外的平台和竹林间长长的台阶或坡道将这两个小房子连为一体。之后拿着草图与老陈又交流了几次，基本确定后，方案的深化和完善是我的研究生戚骁锋完成的，他是一个完美主义者，为了准确地确定建筑各部分的标高、构造做法等折腾了好几个星期。最终当我们把方案拿给老陈的时候，他非常满意。

为了节省资金，建筑的施工是老陈的一个叔叔带领着的一帮“老伙计”干的。老陈夫妇几乎每天都待在工地，妻子负责给工匠师傅准备午餐，老陈更多的时候则是在场地上不断琢磨，看如何优化方案。为了师傅们更好地理解图纸和方案，我和我的学生把模型搬到了现场，大家一起想象着建成完工后的样子。

房子刚开始建的时候，开挖、放线、做基础，一切进展顺利。当钢结构立在场地上的时候，更是显得有模有样了。中间有一段时间我没有去工地，待我再去的时候，惊讶地发现老陈请了一批专做竹建筑的工匠，在现场正紧张地忙碌着。老陈说，要在原来方案的外廊上加盖一个竹结构的顶棚，这样廊子就可以利用起来作为休憩场所了；另外，竹廊的转角做成起翘的形式，这样看起来更有园林和乡土气息。面对如此大的调整，我有点惊讶，不过也理解。

作为建筑师，通常都希望建筑尽可能抽象、纯粹、简洁；作为一个业主，更多时候则希望自己的房子有着某种象征或符号的味道。这种差异我想很大程度上源于角色的不同。对于建筑师而言，房子只是房子；可是对于老陈而言，这个房子是他的事业，除了如何经济合理地建造，其未来的服务目标人群以及运营方式他都必须通盘考虑。我相信老陈的决定是经过了充分思考的，因为这需要一笔不小的额外费用。师傅们的效率很高，不久一个亭台楼阁般的竹建筑就立了起来。

老陈另外一个让我眼前一亮的动作是在场地上增加了一个不大不小的景观水池，那也是他在现场对于场地不断探寻的结果。场地附近有山泉水，他将它们引入到水池以及用陶罐自制的喷泉之中。因为水的映照，房子多了些灵气，整组建筑设施也更像是一个小小的园林，即便没有外在的风景，自身也具有足够的吸引力。

现在回想起来，在走出竹林的那一刻，老陈自己心里也早已经有了方案，他只是没跟我说。不过，说实话，他的方案比我们的要好，因为更加实用，而且每次在现场看

was finished, we took it to Chen and had some discussions before we decided on the final design. One of my graduate students Qi Xiaofeng was the one who completed all the details. He is a perfectionist, so it took him several weeks to determine the exact numbers for heights and structures of the building. The final design turned out to be quite satisfactory to Chen.

To keep the cost within the budget, we hired one of Chen's relatives, his uncle with his own crew to do the actual construction work. Chen and his wife stayed at the construction site almost every day, working side by side with the whole team while providing meals. Mr. Chen never stopped thinking and seeking every possible improvement to the original design. To help the builders better understand the blueprint, my students and I moved the house model to the construction site so that they could have a clearer picture of what the finished building would look like.

The beginning part of the work, laying the foundation, went quite smoothly. After the steel structure got erected on the ground, everything seemed to be on the right track so I stopped going to the site for a few days. However, the next time I went back to find out how things were going, I was surprised to see that Chen has hired a team of craftsmen who specialized in bamboo building. When asked, Chen told me that he decided to add a bamboo ceiling to the verandah which would be able to serve as a resting place for his visitors. Apart from that, the four corners of the verandah ceiling would be in an up-curve shape which looks more rural and garden-like. As surprised as I was, I found such modifications did make sense.

Architects are always looking for the simplest and purest plan, whereas property owners would prefer that their houses can convey some kind of symbol. The difference in priority largely stems from the difference in the roles they play respectively. To me, an architect, a house is a house, nothing more. However, to Chen, this house is his whole career, so he has to take everything into consideration, including not only its cost, but also the potential customers as well as the operating of his business. I believe all his decisions have undergone thorough considerations, including the extra expense for the addition of the bamboo ceiling. Working in high efficiency, these bamboo building masters managed to complete the whole bamboo structure within days.

Another thing Chen did that really impressed me is the addition of a medium-sized water pool which is another result of his constant exploring on the site. There is water flowing down from a mountain spring nearby the construction site and he managed to divert the water into a water pool as well as a fountain made with pottery pitchers. Because of the reflection of water, the building has a little more aura. The whole building facilities are more like a small garden. Even without the external scenery, it itself also has enough attraction.

Looking back now, I realized that, actually, Chen already had a complete plan hidden in his head the moment we walked

到他的朋友们对此房子频频拍照，我就更加明白了他要采用如此材料与形式的原因了。

与一般的项目有所不同，该项目的建设背景是先有产业，后有建筑，而且建筑的出现，又提升了该产业的发展。项目的业主老陈在此承包竹园进行竹酒的生产与加工已经有几个年头，所经营的业务日趋成熟，而且积累了稳定的客户群，也正是在许多客户的建议下，老陈才下定决心修建此设施用房。这一方面可以改善竹酒现场加工的条件，另一方面可以以此作为基地，延伸出相关的竹酒制作、展示、体验、品尝，以及休闲娱乐等活动，从而带来更好的效益。所以在建设之初，老陈就对于该房子的布局、如何使用都有过通盘的谋划，其实，这种"谋划"一直持续到今天，其谋划的过程也是房子的空间与功能不断优化的过程。这组竹酒设施，规模不大，用材与建造也比较简单粗放，但在实际的日常运营与体验中，却呈现出了很强的活力与吸引力，这在很大程度上归功于老陈全过程参与经营。先

out of the bamboo patch. In honesty, I think his plan was better than ours in practicality. Now, every time I see his visiting friends can't stop taking pictures of the house, I will have a better understanding of his choice in certain building materials and structures.

Unlike other projects, the construction background of this project is that there is an industry first, then the building. Moreover, the appearance of building promotes the development of this industry. Before this project got started, Mr. Chen has been working in bamboo-wine brewery for several years and his business has been well-developed with a stable clientele base, which is quite unlike the normal situation where the building project precedes the developing of a business. It was under the advice and suggestions from his customers that he finally decided to build this place and use it as a better equipped wine brewery factory as well as a place for more bamboo-wine-related and benefit-generating activities, such as wine exhibiting, wine-tasting and other entertainment for his customers and visitors. Therefore, at the very beginning of the construction, Chen had made a thorough plan concerning the basic layout and functions of the facilities. In fact, his planning has kept going till today and the whole process of planning

有一个用心的业主，之后才有一个有趣的房子，我想，这是这个建筑取得成功的一个重要原因。

乡村里有不少类似老陈这样的微企业，这些微企业通常只是消耗着村庄的资源与环境，很少与村庄和村民有更加深入的互动与合作，但是在这组竹酒设施里，我们却看到了一种“村企合作”的良好状态。这种状态的出现，一方面是源于老陈与村里的良好关系，另一方面也是双方意识到了合作后共赢的局面。竹酒基地与村民居住区之间，是一大片村集体经营的樱桃园，村里意在发展樱桃采摘、农事体验、亲子活动等项目，但由于刚刚起步，加上樱桃园采摘的季节性极强，所以效益还不明显。随着竹酒设施的出现，村企之间看到了在产品上差异互补、经营上合作发展的可能与潜力，于是将其道路交通、停车场、游客休憩设施等纳入了统一考虑。通过这些设施的建设，村庄、樱桃园、竹酒设施在交通上互通互达，在视线上互为景观，在服务上合作互补，同时，樱桃园、竹酒设施等又纳入了鄣吴镇的全镇旅游系统之中，并成为其中的一个重要的节点，如此的“自下而上”“自上而下”相结合的发展策略与过程，给玉华村的未来发展奠定了良好的基础。

虽然我们是这组竹酒设施用房的最初设计者，但是真正的建筑师其实是业主老陈。建设过程中每次去项目现场，常常会有惊讶，但更多是惊喜：形式的预想与意外、工匠的自在与在意、景观的不变与多变……我们常说建筑师在乡村里进行设计应采用引导而非主导的工作方式，我想这应该是一个很好的案例。

引导而非主导的设计策略，要求建筑师放弃通常的“作品”意识，与业主进行充分沟通，接受场地的各种限制与制约，并适度在材料、空间、形式等多个方面对原设计方案进行调整、妥协，因为在乡村里，相比建筑师，村民和业主无疑更加懂得自己在生产、生活中的需求，村里的工匠也更加知道如何以一种更为经济合理的方式搭建一个空间。建筑师以如此的方式进行设计与建造，建筑在空间、形式上的“作品感”肯定会降低，但是建筑往往会以一种更加符合乡村发展需求的方式融入乡村的环境与日常的运营之中，其被使用后的“鲜活”状态赋予了建筑另一种魅力。对于该竹酒设施而言，业主对于建筑师最初的设计方案，在总图、平面、立面、材料选择、建造方式等方面都进行了较大的调整。总体上，在经过业主的调整之后，建筑变得更加经济实用，特别是获得了比较大的外廊使用面

and revising was also the process of constant optimizing. The whole batch of wine-brewery facilities are small in scale, and simple in material and construction. However, they showed a strong vitality and attraction in the daily operation, which to a great extent is credited to Mr. Chen's participation from the beginning to the end. I believe one of the reasons why this project has been so successful is the participation of a fully dedicated owner.

There are quite a lot of small business just like Chen's in rural villages, but they seldom have any closer interaction as well as cooperation with the local villagers. In contrast, from these bamboo-wine brewery facilities, we have seen a mutual beneficial business model of "business-village cooperation". The emerging of this model originates in part from a good relationship between Mr. Chen and the village, and in part from the fact that both sides have been striving for a win-win solution through their close cooperation. Between the wine brewery base and the village residential area sits a large cherry orchard owned by the whole village. The village council has decided to promote a business project combining cherry-picking, farming experiencing and parents-children bonding activities. However, since the whole project is still at its initial stage and also due to the seasonal feature of cherry-picking, the business hasn't started making profits yet. Now that the bamboo-wine brewery facilities have been built, the possibility and potential of cooperation between the two projects became apparent to the village. Therefore, to make them part of a large-scale business model, the village has decided to put more complementary facilities into their developing projects including building new roads, more parking space, as well as resting and recreational facilities for visitors and tourists. With the addition of these facilities, the village, cherry orchard and bamboo wine brewery will be closely connected in geological positions and also complement each other in the products and services they provide. Meanwhile they would all be included into the whole tourist system of Zhang Wu town and become an important joint connecting two developing strategies, one being "top-down" and the other being "bottom-up". It will lay a good foundation for the future development of Yuhua village.

For these bamboo-wine brewery house, we are said to be its original designer, but in actual fact, the real architect is Mr. Chen, the business owner. Every time I went to the building site during the construction, I was always amazed by the unexpected addition and modifications made by Chen, the initiative and dedication from the builders and craftsmen, as well as the ever-changing landscape, etc. We always talk about the guiding, not dominating working strategy that architects should adopt in rural villages, and I believe this is a great example.

The designing strategy of "guiding instead of dominating" entails that architects should forgo the consciousness of "works" mindset and instead, they are supposed to have plenty of communications with the property owner, being aware of all

积，以及营造出了较强的乡土与园林气息，这对于一般的消费者而言，相较之前的现代简约风格，更加容易接受。

竹酒设施北面的樱桃园，村里去年也进行了整修，增设了步行小道以及停车设施，以方便游客采摘。今年春天，樱桃花开得很漂亮，从竹酒设施的外廊上看过去，让人心旷神怡。经过两三年的建设，玉华村的竹酒、樱桃园、茶园等呈现出联动发展的良好格局，让人对其未来充满了期待。

sorts of limitation and restraints and ready to make appropriate adjustments as well as compromises in the aspects of building materials, space and forms of the original design. You see, in rural areas, villagers and the property owners more often than not know more than architects about what they really need in their daily life and work. Likewise, the local builders and craftsmen normally have a better idea of how to build in a more reasonable and economical way. Following this strategy, although architects' personal signatures in spatial and formal features would be downplayed, the buildings would be able to naturally integrate into the rural environment and the daily operation of the rural life in a way better suited to the rural development. In the case of Chen's bamboo-wine brewery facility, he has made some major revisions and adjustments to the original design in the overall layout, the building materials as well as the building method.

Overall, after the adjustments made by the owner, the building has become more economical and practical, in that

it created a bigger space for the verandah and a strong rural and gardening atmosphere, which makes it easier to be accepted by the customers than the modern simplistic style of the previous designs.

The cherry orchard to the north of the brewery facility has undergone some renovation as well, with a small walking path and a parking lot being added for visitors. The cherry blossoms are especially exuberant and pretty this spring which offers a very pleasing sight for people who look out from the verandah outside the brewery house. After three years' construction, the bamboo-wine brewery, cherry orchard and the tea garden have shown a well-coordinated developing momentum which fills people with anticipation for a better future.

玉华村竹酒设施用房

一层平面图
First Floor Plan

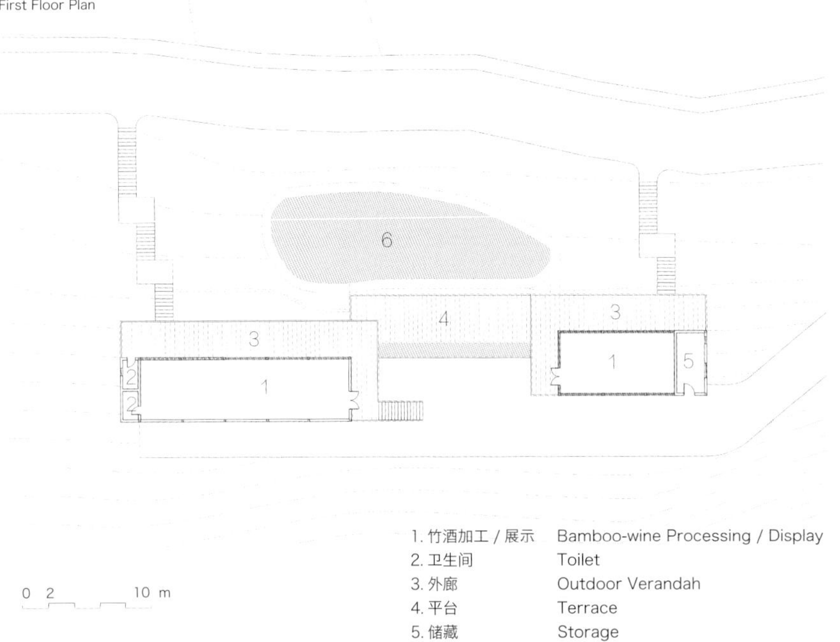

1. 竹酒加工 / 展示 Bamboo-wine Processing / Display
2. 卫生间 Toilet
3. 外廊 Outdoor Verandah
4. 平台 Terrace
5. 储藏 Storage
6. 水池 Pond

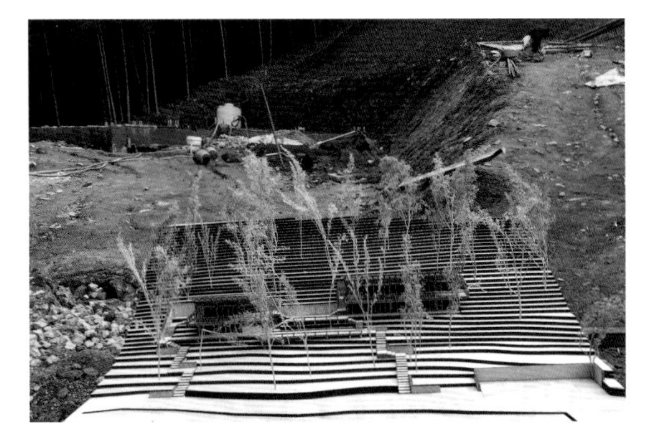

建造过程 I Building Process

鄣吴四厕

Four Public Toilets in Zhangwu Town

安吉县鄣吴镇 | Zhangwu Town, Anji County

厕所位置分布图
Toilets Location Map

1. 昌硕广场厕所 Toilet on Changshuo Square
2. 鄣吴溪沿岸厕所 Toilet on Zhangwu Creekside
3. 八府广场厕所 Toilet on Bafu Plaza
4. 玉华村厕所（具体位置参见 P252~P261） Toilet in Yuhua Village(See P252~P261 for the location)
5. 吴昌硕故居 Former Residence of Wu Changshuo

如果按照类型学来划分，我想厕所应该是一种最古老的建筑了，因为如厕的需要与生俱来，那是来自天性与自然的召唤，正如英文中如厕的诙谐表达：“Nature is calling me.”随着人类文明的进步，厕所从露天进入了房间，但毕竟是“难言之隐”，所以公厕在建设中，往往被放置在场地的边缘角落，通常也不会成为建筑师设计的重点。但是随着社会的发展，人们对于厕所环境的要求越来越高，甚至有人提出了“厕所革命”的口号，近些年来，以旅游部门为主导推出的星级公厕项目，在各地开展得更是如火如荼。人们如此重视，自有其充分的道理。君不闻：“小厕所，大问题”。

厕所的蹲位隔间，通常的尺寸是0.9 m×1.2 m，我想这应该是个人在所使用的建筑类公共设施中最为狭小也最为私密的空间了。在关上隔间门的那一刻，你仿佛暂时逃离了这喧嚣纷繁的世界，成为它绝对的主宰。那个时候，你也许会微闭双眼，休息片刻，或是在手机上读上一篇短文，发出一条消息，总之，那个时候多是安定和自在的，也很有可能是最有创意和想法的，很多非凡的构想和发明据说就诞生在那里。如此想来，厕所的责任和意义很大，其建造与设计来不得慢点马虎，空间序列、光线、景观、材质等都需琢磨，因为厕所，不仅仅是厕所。

我在不少地方，就看到各式各样的别致厕所。有的用料考究，有的形式独特；有的配备了光伏板、能源自足；有的有着大面积的玻璃屋顶，阳光肆意；有的还有音乐在你的耳边响起。我在杭州运河边武林广场段看到一个厕所，其内部居然自带了一个微缩园林。现在的星级厕所都配备了第三卫生间，满足婴幼儿和其父母，以及需要照顾的人群的需要，人性化十足。可是，很多时候，面对一些形式过于夸张、装修过于奢华的厕所，我总是还不太习惯。这一方面源于豪华公厕与周边环境的强烈反差，另一方面则是它们远远超越了人们“如厕”这一基本的功能需求。厕所，做到了干净整洁，或许，只要再讲究一点点就好。因为厕所，毕竟只是厕所。

以下几个公共厕所，是我们近几年在浙江鄣吴镇设计建成的部分项目，有的只是为了满足村民日常生活的需求，有的则是纳入了景区星级公厕的范畴。项目都很小，可是每个项目所在的场地环境、业主要求都不太一样，需要尽全力量身定制，所以其特点也各不相同，如下一一道来。

In a typological term, I believe toilet should be one of the oldest architecture due to fact that "using the toilet" is an inherent and natural need for human beings, hence the humorous expression in English language: "Nature is calling me." With the advancing of human civilization, the location of toilets have been moved from outdoor to indoor. However, since it's such a private matter, public toilets have always been placed in some remote and obscure corners. And they have never been part of the main focus of an architectural design. Nevertheless, with the rapid development of the modern society, people are having increasingly higher standard for toilets in terms of its functions and sanitary condition, which has been called "the toilet revolution". What's more, the tourist industry has started to promote such projects as "the star-rated public toilets", which are all the rage across the country. The fact that so much attention has been given to toilets is not without reasons, as people always say: "little toilet, big problem."

The normal scale for a single stall in the public toilet is 0.9 m×1.2 m, which, I think, is the smallest and the most private space for individual use in the domain of public facilities.The moment you close the cubicle door, you seem to temporarily escape from the hustle and bustle of the world and become its absolute master. At that time, you may close your eyes, take a break, or read a short article on your mobile phone and send out a message. In shot, you will be more calm and comfortable at that time. Moreover, it might be during the toilet time that some of your most creative and original ideas come to you. It was said that a lot of exceptional inventions were born in a toilet. Therefore, I believe it is safe to say that there's great significance and responsibility in the designing and constructing a toilet. So many factors and elements need to be thoroughly considered, such as the spatial order, lighting, materials as well as decoration.

I have seen various types of public toilets in lots of places. One particular toilet I saw at Wulin Square along the bank of the Grand Canal in Hangzhou even contains a mini-garden. Every "star toilet" also provides a special room for parents with their infant kids and those in need of care, which is a very humanistic design. With all that's been said, I, personally, still find it hard to understand the existence of those "luxurious" public toilets with extravagant decoration and abnormal even bizarre shape. They are either in stark contrast with the surrounding environment or unnecessarily above its most basic function, a place for people to relieve themselves. The way I see it, a toilet is a toilet, so being clean and neat with a touch of ornament is good enough.

The four public toilets that are going to be discussed in the following are part of the designing projects we had finished in Zhangwu Town, Zhejiang Province. Among these four projects, some are simply for local people's everyday use, while others belong to the group of those fancy, star-rated public lavatories. Although they are all relatively small projects, their respective location, surrounding conditions and requirements from the owners are completely different, therefore, customization for each one of them is necessary. Here are the details.

1. 昌硕广场厕所：空间上的织补

建筑师： 贺勇，王凯伦

项目地点： 浙江省安吉县鄣吴镇昌硕广场边

建筑面积： 84 m^2

建成时间： 2017 年 12 月

1、Toilet on Changshuo Square: Space Weaving

Architects: He Yong, Wang Kailun

Site: Changshuo Square Zhangwu Town

Gross Floor Area: 84 m^2

Date of Completion: December 2017

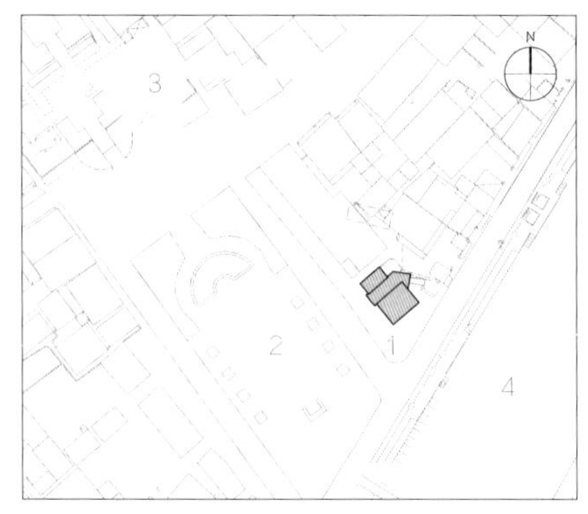

1. 公厕 Public Toilet 2. 昌硕广场 Changshuo Square
3. 吴昌硕故居 Former Residence of Wu Changshuo
4. 鄣吴溪 Zhangwu Creek

总平面图 Site Plan

场地是昌硕广场周边余下的唯一一处狭小的三角地，紧邻其边缘已经有了一个公交自行车棚、一个高杆广场灯、一棵大树，这些都需要保留与避让，所以场地显得更加局促。周边的邻居担心厕所的味道，要求蹲位离他们的宅院要尽可能远，而且临近他们家的墙面不要开窗，再加上要满足三星级公厕的一系列要求，如此下来，此项目的设计条件可谓苛刻之极。

根据场地的分析，我们发现从道路和广场都最好能有进入公厕的可能，于是巧妙地借用了邻居家的院墙，创造了一个穿越该公厕的“街巷”，洗手盆就放置在“街巷”中间，从“街巷”走过，你会看到“微型广场”上的青松、扇形窗洞外的翠竹，以及月亮门外的广场景观，在这里，“如厕”变成了一个漫游的过程。

与其说这是一项设计，不如说是一道空间上的完形填空。受制于场地的狭小以及传统村落风貌的要求，该建筑的布局、样式、色彩在很大程度上都早已经被决定了，建筑师要做的是如何以一种非常克制的态度与方式，将建筑合宜地置入场地之中，使其虽然新建，可似乎原本就存在于那里。

Its location is on a small triangle-shaped ground at the edge of Changshuo Square. Since its surrounding area had already been occupied by a public bicycle shelter, a high mast lighting, and a giant tree, which all need to be kept as they were, what's left for the new project is a very limited space. Besides, people living in the same area all expressed their concerns about the possible bad smell which might emanates from the toilets so there could not be any windows in the toilet wall adjacent to their houses. And above all, this particular project was supposed to meet the standards for a three-star public toilet. Taking all the factors and requirements mentioned above into consideration, it was truly a very demanding job.

Based on an overall analysis of the location, we believe it's necessary for the toilet to have two entrances, one facing the square and the other opening to the road, so we came up with this "clever" idea of creating a small "alley" which runs through the toilet building. A section of the neighbors' walls bounding their courtyards will automatically become one side of the alley wall. The handwash basin has been placed in the mid-section of the "alley", walking through which you would be able to look through the fan-shaped window and see the evergreen pine tree, the verdant bamboo as well as the beautiful scenery of the square.

For the architect, this designing work was more like a cloze question in an exam. Since the layout, style and the color of the new building supposed to adopt had already been decided, what the architect needed to do is fit the building into its location in a controlled and restrained manner, creating the illusion that this public toilet has been there all along instead of being newly-built.

昌硕广场厕所

鸟瞰图
Bird's Eye View

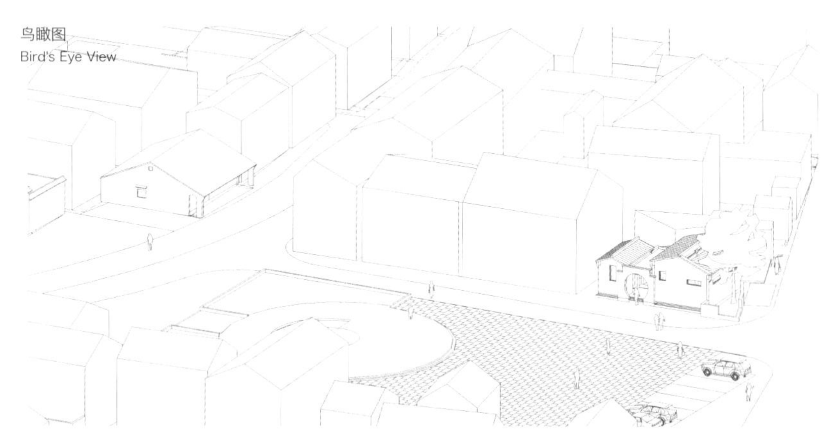

一层平面图
First Floor Plan

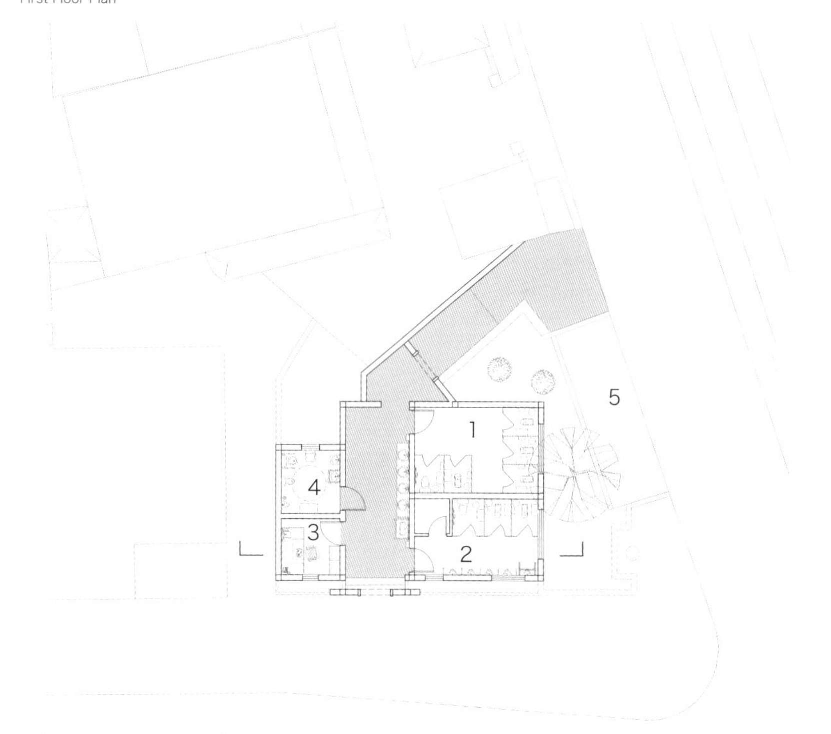

1. 女厕 2. 男厕 3. 管理间 4. 第三卫生间 5. 自行车租赁点
1. Female Toilet 2. Male Toilet 3. Management Room 4. Third Toilet 5. Bicycle Rental Point

剖面
Section

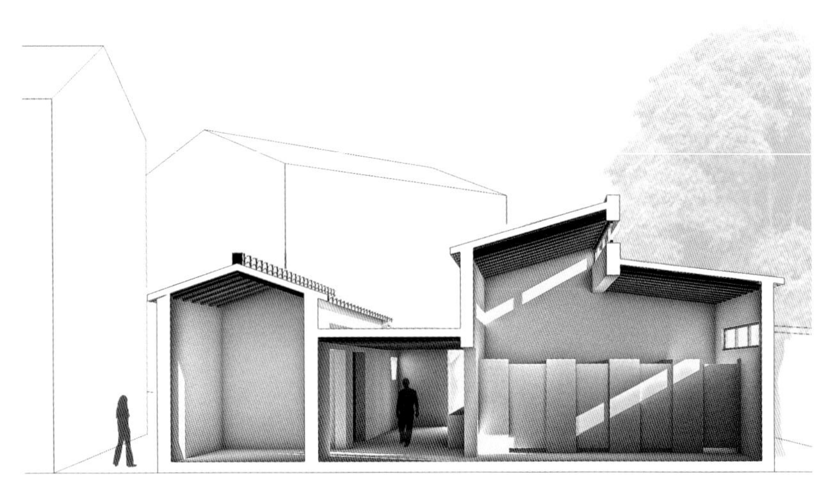

昌硕广场厕所

2. 鄣吴溪沿岸厕所：边界交汇处的自在
建筑师： 贺勇，陈耀
项目地点： 浙江省安吉县鄣吴镇鄣吴溪边
建筑面积： 112 m²
建成时间： 2017 年 12 月

2、Toilet on Zhangwu Creekside:
A Spot of Ease at the Edge of the Boundaries
Architects: He Yong, Chen Yao
Site: Zhangwu Creekside, Zhangwu Town
Gross Floor Area: 112 m²
Date of Completion: December 2017

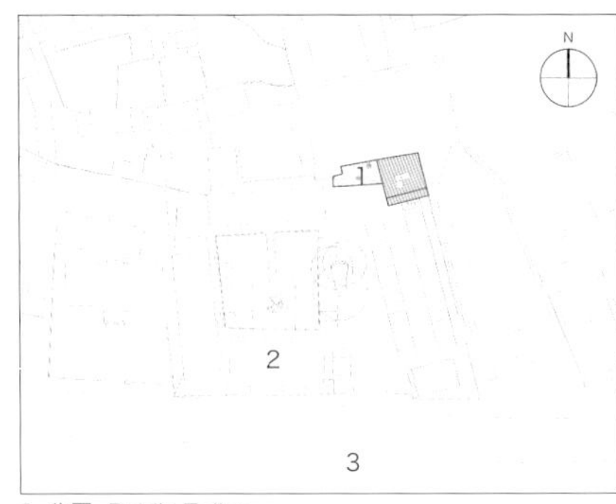

1. 公厕 Public Toilet
2. 鄣吴镇政府办公楼 Government Office Building
3. 鄣吴溪 Zhangwu Creek

总平面图 Site Plan

该项目是鄣吴溪沿线景观整治工程的一部分，位于政府大院内的广场上。那里原本有一个公厕，但由于建设年代过于久远，已经非常破败，而且位居场地比较居中的位置，既影响了景观，也不利于场地停车功能的组织，所以亟需结合鄣吴溪沿线景观提升工程，进行整体的规划与建设。

因为要满足大量停车的需要以及未来各种较大型群众文娱活动的弹性需求，广场上除了保护好那几棵大树，不能有其他过多的介入。于是只能在边界上做文章：设计中顺着原来的东侧围墙布置了长廊，作为日常宣传展示的地方；北侧的围墙不太规整，无法满足国旗杆背景墙的要求，于是在那里用高墙围合出了一个小小的花园，既塑造出了一个干净的背景界面，又在原本单调的广场上创造了一个幽静的空间。如此下来，两个边界交汇处的所在，自然就是公厕的最佳位置。长廊、公厕、小花园也在广场周边共同构成了一个富有趣味的漫游环线。

公厕是简洁的白色方盒子，其东边有几株高大的香樟，枝叶繁茂。在绿树的掩映下，白色的盒子愈发明亮，只是室内的光线差了些，于是在白色盒子顶部的中间开了几处天窗，分别将男、女卫生间的洗手处照得温暖明亮。

This project was part of a regulation campaign directed to the landscape along the bank of Zhangwu creek. The toilet is located in the inner square adjacent to the government building. There used to be an old public toilet before but it had become quite dilapidated after years of wear and tear. Being at a center position, it had been seen as a smear to the overall landscape and it also got in the way of the functioning of the parking service in that area. That's why a general re-design and re-construction is urgently needed in order to improve the overall landscape along the bank of Zhangwu creek.

In order to meet the elastic demand for car-parking as well as public activities of a large scale, no drastic alteration or interference allowed to the setting of the square. Even those tall trees are to stay exactly where they were. The only place for us to make any change is at the edge of the square. In our design, a long corridor will be placed along the bounding wall to the east side of the square, which also serves as a spot for daily publicity display. The northern side of the bounding wall has some sort of zigzagging shape so we have circled out a small gardening area by building another section of curving wall, creating a space of serenity and secrecy. This way, the converging spot between these two walls will naturally be the perfect place for the public toilet. Meanwhile, a more intriguing route will be created by the relative positions of the corridor, the small garden, the public toilet and the edge of the square.

The actual toilet is in the shape of a simple white cube, with a number of sky-reaching luxuriant camphor trees. The white color of the building particularly stands out against the background of green trees. The only problem is that it won't be bright enough inside the building. That's where these skylights on ceiling come in which allow the natural lights rushing through, lighting up the whole toileting area as well as the hand-washing counters.

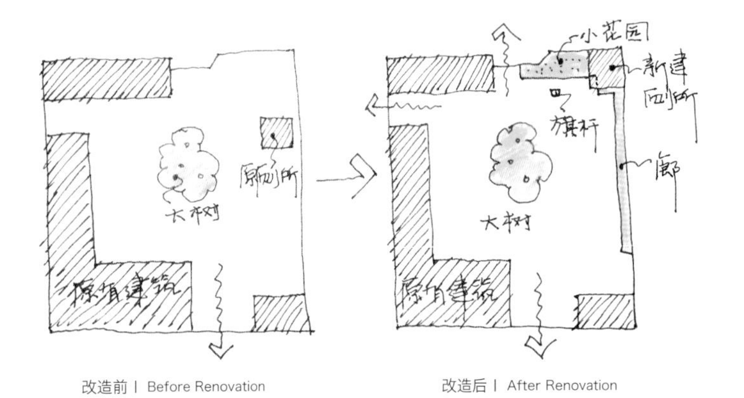

改造前 | Before Renovation　　　　改造后 | After Renovation

场地原状 | Original Status of Site

场地现状 | Present Status of Site

鄣吴溪沿岸厕所

一层平面图
First Floor Plan

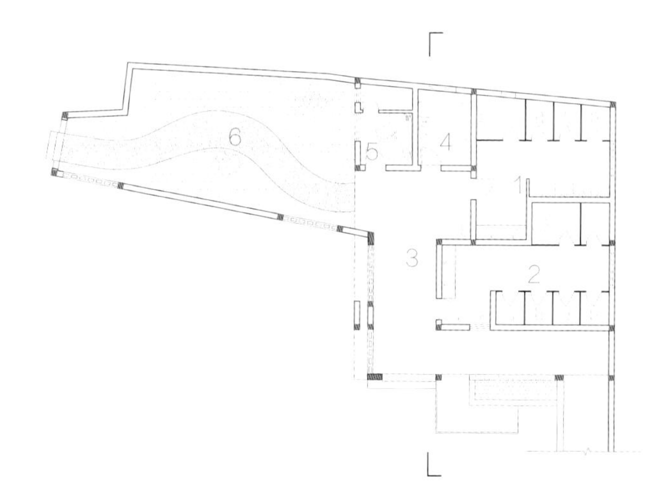

1. 男厕 2. 女厕 3. 前厅 4. 第三卫生间 5. 管理间 6. 院子
1. Male Toilet 2. Female Toilet 3. Front Hall 4. Third Toilet 5. Management Room 6. Yard

剖面
Section

3. 八府广场厕所：树林背后的光影
建筑师：贺勇，孙姣姣
项目地点：浙江省安吉县鄣吴村八府广场
建筑面积：90 m^2
建成时间：2015 年 12 月

3. Toilet on Bafu Plaza: Light and Shadow Hidden in the Woods
Architects: He Yong, Sun Jiaojiao
Site: Bafu Plaza, Zhangwu Village, Zhangwu Town
Gross Floor Area: 90 m^2
Date of Completion: December 2015

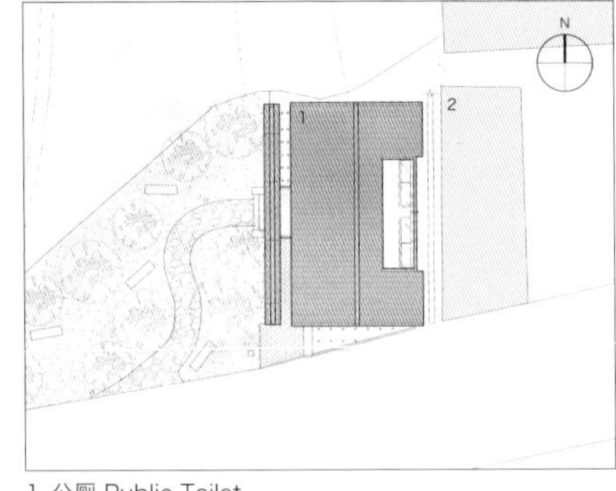

1. 公厕 Public Toilet
2. 现状农居 Existed House

总平面图 Site Plan

该厕所位于鄣吴村八府广场的东南角。场地上有一丛长得极好的香樟苗木，村里原本计划把这些苗木移走，然后建设公厕。在我们的建议之下，村里保留了苗木，并说服了旁边的那户人家，拆掉了部分搭建的棚屋，腾出了空间作为公厕建设之地。现在回想起来，这应该是村里所做的最为英明的决策之一。

得益于那片香樟林的掩映，无论什么样的形体放在那里都不会太难看。不过，如何让建筑与周边的环境更好地协调，还是颇费脑筋的。经过多方案的比较，将公厕的山墙面沿着广场布置，不开窗洞，如此下来，既让厕所获得了私密，又让其形体与相邻的宅院融为一体，仿佛一体规划与建设一般。

如果你现在去八府广场的公厕，穿过那丛茂密的樟树林，或在一片白墙的引导下，进入小门厅后首先看到的是一片镂空瓦片墙，其背后隐约透露出光亮，那里是充满了阳光的洗手台。随后进入如厕空间，那里有一个条带形的窗，窗外是如画的田园风景；还有一片落地磨砂玻璃，竹子在其上投下斑驳的剪影。

This toilet is located at the southeast corner of the Bafu Square in Zhangwu village. There is a patch of camphor seedlings which are growing really well on the site. The village planned to move these seedlings somewhere else before the construction begins. However, under our suggestion, the trees got to remain where they were. We even persuaded the family living next door to tear down part of their shack giving us more space for the public toilet. In retrospect, that may well be one of the wisest decisions made by the village authority.

Under the cover of those camphor trees, building of any shape will look slightly better than it really does. However, there's still a lot of brain-work needed to make the new building fit into the surroundings seamlessly. After comparing multiple schemes, we decided to set the gable wall along with the boundary of the square with no windows on it which both gives the toilet privacy and makes it merge into the neighboring residence.

To get to the public toilet at Bafu Square, you need to either walk through that thick woods of camphor trees or walk along a white wall and into the small hallway where what come into your view is a section of perforated tile wall with hazy lights shining through from the hand-wash counter. After that, you would reach the toileting area where there is a ribbon-shaped window through which you can see the pastoral scenery outside. There is also a French window made of frosted glass with the mottled shadow cast by the bamboo trees.

八府广场厕所

设计构思 | Design Concept

场地原状 | Original Status of Site

设计方案 | Design

一层平面图
First Floor Plan

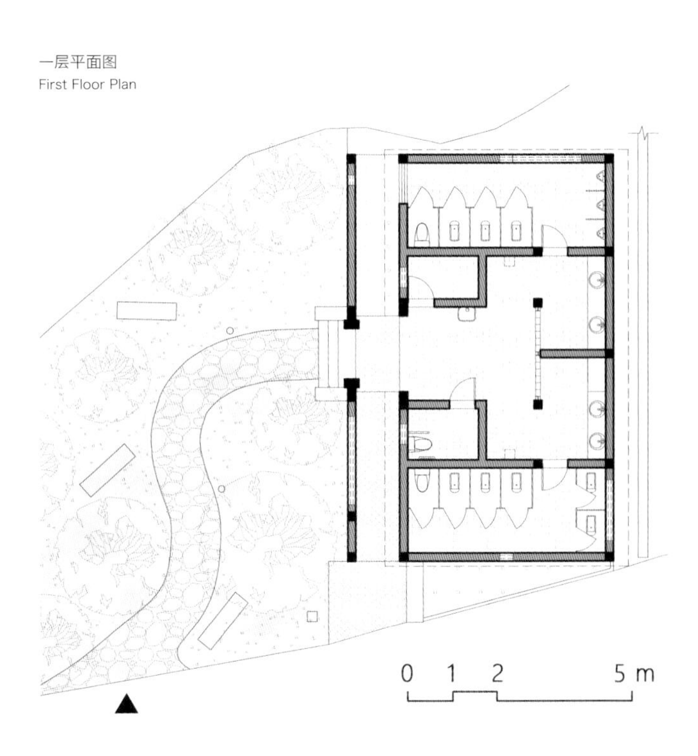

剖面
Section

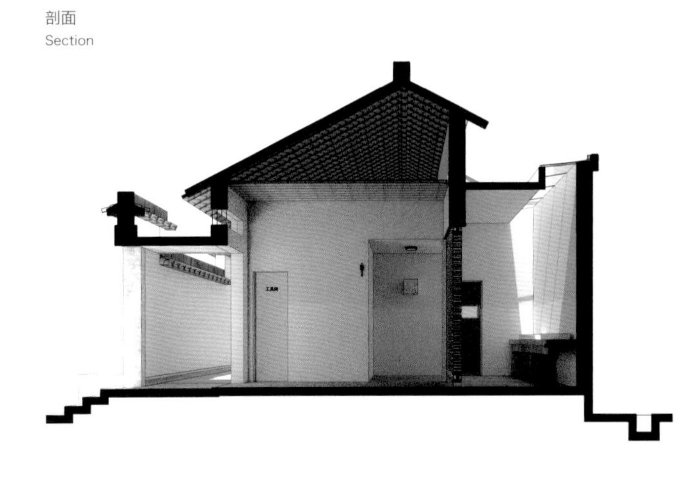

4. 玉华村厕所：融入场地的景观
建筑师： 贺勇，朱博
项目地点： 浙江省安吉县鄣吴镇玉华村口
建筑面积： 68 m^2
建成时间： 2014 年 4 月

4. Toilet in Yuhua Village: Landscape into the Surrounding
Architects: He Yong, Zhu Bo
Site: Yuhua Village, Zhangwu Town
Gross Floor Area: 68 m^2
Date of Completion: April 2014

1. 厕所 Toilet 2. 停车场 Parking Lot 3. 茶园 Tea Garden
4. 水塘 Pond 5. 小河 Small River 6. 农田 Farm Land

总平面图 Site Plan

该公厕位于玉华村口，来往的村民和游客都要路过此地。在这样一个位置修建厕所，建筑体量既要彰显又要适当隐藏：彰显是出于让路人容易发现，隐藏是基地场所环境的要求。所以在布局上，将通常一个完整体量的公厕拆分成三个小尺度的三角形的功能单元（男厕、女厕、洗手处）隐藏在绿化之中，然后用一片较长的清水红砖弧形墙体连接起来，给予使用者较为明确的引导。

从一片延伸至地面的屋顶覆盖下进入洗手处，由此进入两侧的如厕空间需要穿过一个开敞的连廊，这是一个由红砖砌筑的镂空墙体和 PVC 雨水管脱模而成的柱子限定出的空间，其顶棚是清水混凝土，印刻着当地产竹席的纹理。其屋顶选用了深灰色片岩，色泽沉稳且变化丰富，三角形的外墙部分则以卵石贴面，形成质感变化丰富的整体效果。

建筑完工后，其门口原本积满了淤泥的水塘也给予了清淤，在维持其自然的形态之上，对其边坡稍加整治，并种植了些水生植物。不到两年的时间，各种植物长得异常茂盛，也将公厕这一建筑牢牢地锚固在场地之上。

This public toilet is located at the entrance of Yuhua village, a place every villager will pass by every day. The very location of the new building required both prominence which means the building is supposed to be found easily and also concealment due to the demand of its surrounding. That's why, in terms of layout, we decided to divide the whole building into three smaller sized function units (respectively for women, men and handwash area), each in the shape of a triangle. All three parts were to be hidden in green plants and connected with each other by a long curve wall made of red bricks which offers an explicit guidance to the toilet users.

To get to the actual toileting area, one has to walk through an open corridor which is formed by the perforated walls of red bricks and the columns made out of a PVC rainwater pipe mould. Its ceiling is of fair-faced concrete with a special marking characterized by a type of local bamboo mat. The material used on roof is dark grey schist which gives out a sedate yet flowing luster. The outer wall is covered with cobble stones creating a tactile appeal.

After the construction was finished, the pond previously silted up by mud returned to limpidity. We also planted some aquatic plants and tidied up its bank a little without disturbing its natural condition. Within two years, these plants have flourished rapidly and managed to firmly anchored this building on the ground.

玉华村厕所

一层平面图

First Floor Plan

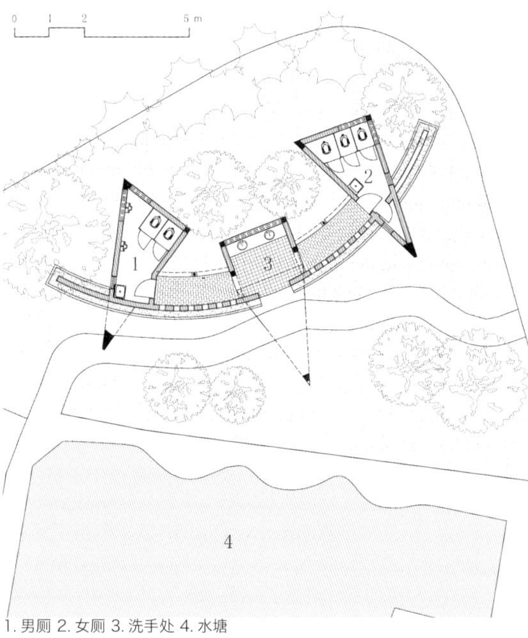

1. 男厕 2. 女厕 3. 洗手处 4. 水塘

1. Male Toilet 2. Female Toilet 3. Handwash Counter 4. Pond

日本建筑师隈研吾在其《负建筑》一书中曾叩问自己：“有没有可能建造一种既不刻意追求象征意义，又不刻意追求视觉需求的建筑呢？”我想，这一系列的厕所应该正好就是这样的建筑，因为在这些小房子的设计过程之中，重要的其实都不是房子本身，而是关系的探讨：建筑与场地，空间与行为，材料与建造。当然，更不会忘了其建设的初衷：作为一项基本、有品质的公共服务设施，满足人们日常生活的需求。

Japanese architect Kengo Kuma once asked himself a question in his book Defeated Architecture. "Is it possible to design a building that neither has to chase after any symbolic meaning, nor need to meet any requirements in visual effect?" To me, this series of buildings are just what he's talking about, because during the process of the designing, what really matters was not the house itself, but the exploration into relationships between buildings and their locations, space and behaviors, or materials and construction. Naturally, we will not forget their primary purpose and that is a basic yet quality public service facility to satisfy people's daily needs.

Yuhua Village Pedestrian Bridge

郭吴镇玉华村 | Yuhua Village, Zhangwu Town

建筑师： 贺勇，葛亚博，浙江中用市政园林设计股份有限公司（施工图）
结构形式： 单腿斜撑（钢筋混凝土）
建成时间： 2018 年 12 月
Architects: He Yong, Ge Yabo, Zhejiang Zhongyong Municipal & Garden Designing Institute
Structure: Diagonal Bracing (Reinforced Concrete)
Date of Completion: December 2018

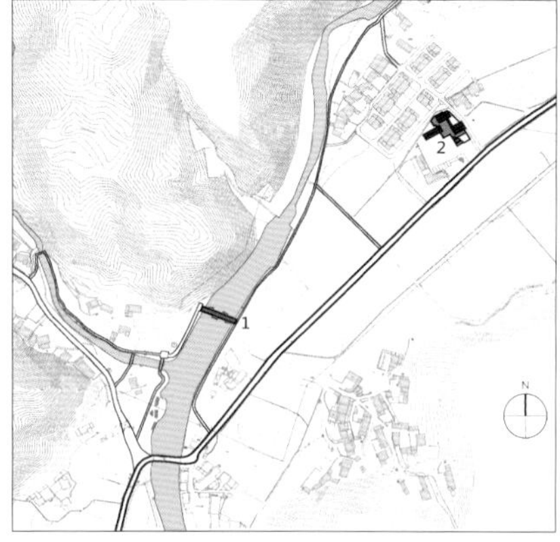

1. 玉华村人行桥 Yuhua Village Pedestrian Bridge
2. 玉华村村委会 Community Center of Yuhua Village

总平面图 Site Plan

缘起

玉华村由几个带状、团状的农居组团构成，各部分相对独立，只能依靠那条穿越全镇的交通干道进行连接，特别是其中的龙亭坞组团，深藏于村西北的一个山谷里，从村口几乎看不到。更加不利的是，那条交通干道在玉华村里通过一座桥跨越鄣吴溪时，转了一个近90度的大弯，桥面很窄，仅能一辆车单向通过，所以那个车行桥往往成为一个交通瓶颈，也是村里一个潜在的危险地方。

近几年，鄣吴镇政府沿着鄣吴溪修建了绿道，这是一个供人步行、骑自行车的慢行道路设施，尽管鄣吴溪沿线风景优美，但是由于它与几个居住组团的联系不够紧密，所以使用的人并不是很多。于是，如何对村里的交通进一步的梳理，使其更加安全、顺畅，加强各居住组团的联系，成了一个迫切需要解决的问题。

1. 穴位处的连接

如果用一个圆把这几个农居组团圈在一起，那么现在所建的玉华人行桥的位置基本就是这个圆的中心。在设计之初，经过仔细的分析，如果建设这座桥，既可以做到人车分流，也能增强村庄各部分的可达性、环通性，特别是

The origin

Yuhua village consists of several clusters of residence areas far away from each other which are only connected by a main road traversing the whole town. In particular, the Longtingwu cluster lies deep in a valley in the northwest of village, which can hardly be seen from the entrance of the village. What's more unfortunate is that the main road takes a 90-degree turn when crossing a bridge over a section of Zhangwu creek within Yuhua village. The bridge is so narrow that vehicles can only cross it one at a time, which was always a traffic bottleneck where all the possible hazards are hidden.

In recent years, a greenway has been built along the Zhangwu creek. It is a pedestrian-only road with a speed-limit for bicycles. However, it has been scarcely used due to its remoteness away from other residence clusters. Therefore, a problem in urgent need of solving for the village would be to strengthen its geographical connections with other living areas and make sure it runs smoothly and safely.

1. The connection at "the acupoint"

If we use a circle to circle these groups together, the new bridge is located at the very center of this circle. At the beginning of design, after a thorough analysis, we decided that the bridge is supposed to carry out multiple functions. Not only should it provide two separate lanes for vehicles and pedestrians, it should also be able to improve the overall

龙亭坞组团与其他各部分的联系，使其到村委会、镇中心都变得更加便捷。若再辅助一个低矮的滚水坝，那么在桥附近就会形成一个开阔的河道水面，极具景观价值，桥以及河道周边也将会是一个潜在的公共休闲场所。

于镇域层面的绿道而言，这座桥的建设，也使得原绿道延伸到龙亭坞组团，让人们可以欣赏到龙亭坞里面幽静的溪流与竹林风景。在乡村多年的调研与实践，我们一直强调“点激活、微更新”，那么桥所在的位置，就是以针灸的方式，解决玉华村目前在互联、通达问题上的关键“穴位”。此处的河道两侧如果得以联通，原本单一的线性绿道将变成一个真正多层级的网络系统。更有趣的是，它将改变村庄的空间结构，让原本离散的农居组团变成一个以水面景观作为中心的紧凑格局。

2. 眺望出的景观

桥的主要目的当然是让人跨越河道，但是桥的意义绝

accessibility of the village as well as the connections among the residence groups. In particular, the connection between Longtingwu cluster and other parts will make it more convenient to reach the village committee and the town center. In addition, It will be complemented by a low overflow dam which would create a naturally formed river channel filled with clean water serving a scenery value. And above all, the area around the bridge and the river channel would become a potential public leisure spot.

Besides, the new bridge will extend the original greenway to Longtingwu, an adjacent residence group which offers a beautiful scenery of bamboo grove and secluded and peaceful stream. We've always been following the strategy of "point activation to micro-update" over years of research and practice in rural areas. In this case, the location of the bridge would be at that vital point in need of activation in order to fix the lack of interconnections for Yuhua village. With the building of the new bridge, the formerly single-way linear greenway would turn to a real multilevel network. What's more, it will change the spatial structure of the whole village by transforming its somewhat isolated and loose structure into that of a compact pattern centered around the water landscape.

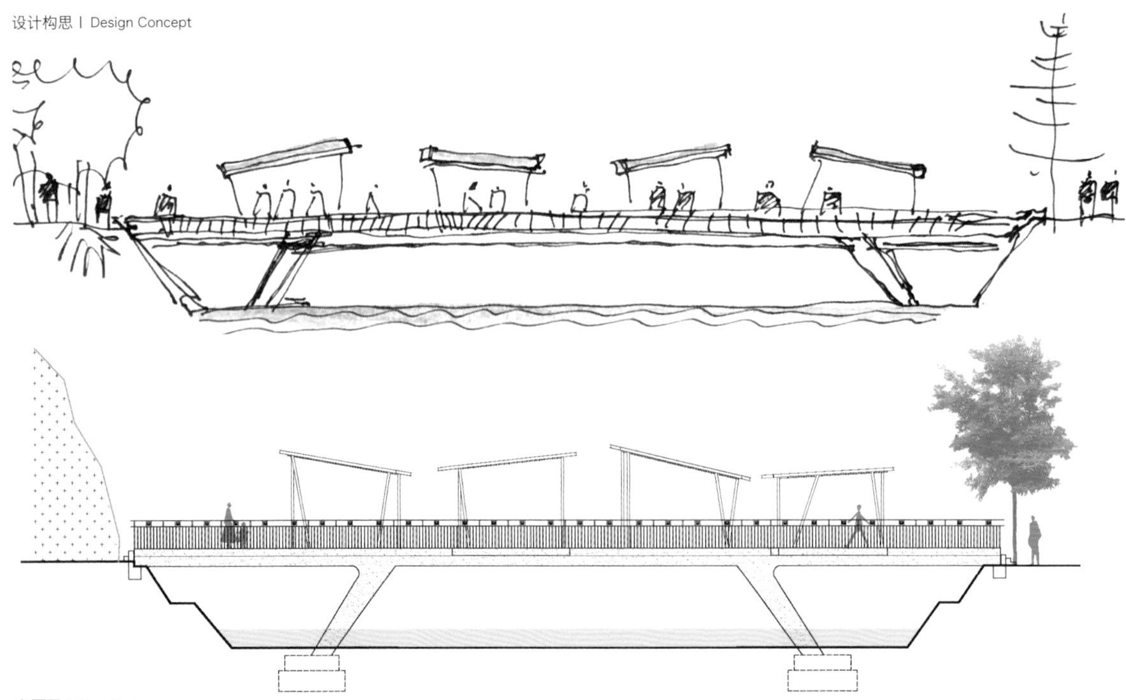

设计构思 | Design Concept

立面图 | Elevation

不仅限于此。正如卞之琳的诗《断章》里所描写的那样："你站在桥上看风景，看风景的人在楼上看你。明月装饰了你的窗子，你装饰了别人的梦。"看与被看、互为风景是人世间最美好的事情，只是我们常常忘了停下来欣赏身边的那些美好，或者没有意识到自己成为他者的美好。鄣吴溪的水声潺潺，一边是郁郁葱葱的竹海，另一边是精心耕作的田园，均是如画的风景，从任何方向看出去，都会投射为你心底的景观。于是，如何让人可以在桥上停留片刻，驻足远眺，眺望出自己那一瞬间的心情，也成就他人的一个微小世界，成为设计之初的一个基本发想和目标。

最初的构思其实就是两个墩子支撑起一座微微拱起的桥面，这样与两侧的堤岸可以更平顺地连接，同时，桥面坡度的变化让人的视线可以微微仰视或起伏，发现别样的风景，但是坡度的出现总让人感觉在桥上的停留不是那么安稳。另外，讨论过程中业主提醒，出于每年秋冬砍伐毛竹季节时运输的安全和方便，希望桥面尽可能平整，同时与两侧的堤岸有一定的高差，这样迫使骑电瓶车的人降低速度，从而保证安全。于是，桥面的形式又回到平直的状态，但是增加了四个出挑的平台，它们既是眺望风景的地方，也是路人避过毛竹运输时的"港湾"。平台是有顶棚的，透明阳光板配合着竹子的吊顶格栅，为驻足在此的人遮挡出一片阴凉。四个顶棚高低起伏，映衬着远处山脉的走势；支撑顶棚的柱子也是歪歪斜斜的，呼应着顶棚的节奏和基调。如此下来，原本单调的桥面有了些轻松、活泼的氛围。

3. 相遇时的惊喜

走在一座人行桥上，那种体验总是令人愉悦的。我常常琢磨究竟为什么会这样，我想在桥上很多出其不意的相遇，应该是其中的一个重要缘由吧。当你来到玉华村人行桥，上桥转弯的那一刻，近在咫尺的茂密竹林或不远处的高大山脉，是桥面上强烈的对景，让你无法忽略；走到桥面中间，不管你是驻足片刻或继续前行，总是会多看几眼河道上的水面风景，映入眼帘的是枯黄的芦苇、悠闲的鸭子。当然，更为重要的，桥上往往是会遇到行人的，如果恰巧是一个熟悉的老友，很自然地闲聊几句则是再惬意不过的事情了。

桥两端与两侧堤岸的衔接空间其实并不是特别顺畅，因为其场地都很局促，而且有九十度角的转折。或许现实就是这样，很难那么完美，只能接受，但设计中依然在桥

2. The landscape seen from afar

The main purpose of a bridge is for people to cross rivers, and yet its significance is way beyond this. It's like what Bian Zhilin, the famous Chinese translator said in one of his poem Part of Article: "As you are enjoying the scenery on a bridge, upstairs on a tower people are watching you; The bright moon adorns your window, but you adorn others' dream." It's so poetic for two people to be a charming sight in each other's eyes. Zhangwu creek is flanked by the eye-soothing lush bamboo leaves on one side and the fertile rice paddy field on the other, which create a picturesque sighting for people watching from all directions. Therefore, the original purpose and aspiration of this project is to provide a special spot for people to stand still and admire the natural beauty which instantly put them into a pleasant mood.

Our first design for the new bridge is a slightly arched surface supported by two piers. However, in our discussion with the property owner, he expressed the necessity of a flat bridge surface for the sake of safety and convenience especially during the bamboo harvesting season. Another suggestion is that there should be a certain height difference between the bridge surface and the river banks, which would compel the e-bike riders to lower their speed out of the safety consideration. Both suggestions have been duly obliged. The new design has a flat surface plus four heightened platforms on it. The platforms come with ceilings, each of which comprises transparent sunray baffle on top of a bamboo grid providing a cool shade for pedestrians. The ceilings have been purposefully set in various heights and the columns supporting those ceilings are accordingly in different lengths. This design creates a lively and casual effect for the otherwise monotonous bridge surface.

3. Surprises at encounters

Walking on the bridge has always being a pleasant experience. Why? I think it's because it offers you plenty of chance encounters, with the eye-pleasing dense bamboo groves in proximity and a backdrop of magnificent mountain range in the distance, with the serene river partially covered by wavy reed and floating ducks, and above all, with an old friend walking towards you from the opposite direction which would naturally lead to a leisurely chat.

Due to the cramped space as well as the 90-degree turn, the connection between the bridge and the river banks at both ends are by no means seamless. However, nothing in life is perfect. The only thing we can do is trying to make the joints look as natural as possible by applying various remedial measures. Another imperfection involves a sewage treatment station to which two modification have been done. First, we have a row of bamboo trees planted in front of it which gives the station some sort of coverage. Second, a new platform has been added to the all-too-narrow walkway along the river bank. Seeing from the opposite side of the river, the new platform

两端节点处，通过适度筑堤、挖山，使得空间局部放大，让衔接不至于过于生硬。桥头附近有个污水处理站房，不可移动，只好顺着其外墙种植了些竹子，稍加遮挡。站房边的原有临水步道过于狭窄，于是在这里设置了一个平台，出挑到河面之上，从河道对岸看过来，这个平台与桥面上伸出的凉亭似乎融为一体，有了一种意想不到的统一和趣味，这种结局于建筑师而言也是一种惊喜吧。

4. 建造中的真诚

桥的跨度接近40米，如果再采用传统的石头拱券结构，在当下人工昂贵的条件下，将是极其费钱的事情，所以桥的材料采用了钢筋混凝土，施工起来方便，又具有一定的容错能力，适应乡村“低技”的施工与建造方式。结构形式为单腿斜撑，桥墩与桥面在结构上融为一体，如此下来，可以使得结构尽可能轻巧；为了进一步减轻桥的自重，在浇筑桥面的混凝土之前，埋入了数根 PVC 的雨水管。

桥作为村里的一项大工程，施工当然是有资质的单位。不过，其中不少的工匠师傅都是本村或者邻村的，他们每天早上六点开工，十一点收工，中午回家或在附近的小饭馆吃饭。下午十二点半开始，一直干到五点，碰上下雨天则休息，如此的工作方式，不紧不慢，应和着乡村的节奏。工程现场负责人王师傅会经常拍几段视频发在朋友圈中，而且一边拍摄，一边讲解着工程的进展，语气里总是充满了自豪。因为各种各样的原因，工地上一些做法的变更，甚至返工也是不少见的，但是师傅们都以极大的耐心顺利地完成了任务，几乎没有什么怨言，这应该是乡村熟人社会才会有的和谐的甲乙方关系吧。当桥面上的棚子立起来的时候，几根支撑的柱子“东倒西歪”，我问几个师傅，你们觉得这个好看吗，他们说这样挺有趣的，不然就太古板单调了。我听后心里甚是窃喜。我相信他们所说的是真话，因为从其语气和眼神，我看到了其中的真诚。

5. 结语

这个方案的落成也是各种机缘巧合的结果。这座桥的设计方案其实早就好了，但是讨论过程中，地方领导希望能有更多更好的方案，于是请其他机构设计了多种样子的，譬如拱桥、廊桥、折线形桥，建筑师们也都是用尽了心思，不过在我的心底始终认为我们的这个方案是最好的，因为它经济、直接，又有那么一点惊喜，我们坚信这才是日常

seems to be right next to the pavilion on the bridge surface which has inadvertently created an effect of unity.

4. Sincerity in construction

Since the bridge spans up to nearly 40 meters, it would cost too much money if we use stone as the main building material, so instead we choose reinforced concrete for its reasonable price. Besides this choice is in accordance with the "low-tech" construction approach characteristic of rural areas. In addition, the single leg diagonal bracing structure as well as the bridge piers are integrated into each other which render the whole structure as light as it can be. To further lighten the weight of the whole bridge, several PVC storm sewers are buried beneath the bridge surface which would divert the rain water into the river almost instantly.

As a "big" project for the village, the hired construction company is naturally officially certified. What's more, a lot of craftsmen of the building team are local people living in Yuhua village or the one next to Yuhua who hardly need any time on commuting, so they are able to work at a quite regular pace every day. Our craftsman masters are fairly conscientious workers who treat their work with great patience as well as pride, holding up to a high professional standard. That's why I really trust and cherish any of the professional advice and comments that I could get from them.

5. Epilogue

The completion of our design for the bridge is a result of all sorts of coincidence and opportunities. At the very beginning, our design didn't get a total approval from the authorities who expected more options. They actually committed other studios in search of other "fancier" or "glamorous" designs. Deep down I just know that our design is the best one because it's the most economical and straightforward plan with a pinch of surprises which, I believe is supposed to be what are really needed in everyday life. Sure enough, after a while, our plan beat all the competitors and got picked.It seems that good things will come true sooner or later .All you need is to wait.

Among all the projects in Zhangwu Town, this bridge is the most significant and valuable one to me. Part of the reason is that it is at just the right location, which has completely reshaped the spatial structure and landscape for the whole village and in turn greatly improved the connection between the village and its surrounding areas. Moreover, it may be because we each year in our hearts for a bridge over which we can find the way to ourselves.

While the bridge is under construction, the supplementary facilities i.e. the river bank walkway and nightscape lighting are being built at the same time. When all the construction was completed, this spot immediately became the most attractive and lively public area. Everyday as the night falls, the bridge area gets lightened up whose glowing reflection on the river

生活所真正需要的。果然经过一段时间，我们的方案又再次成为实施的方案。看来，好的东西迟早是会实现的，你所需要的只是等待。

在郭吴镇完成的众多项目中，这个桥于我个人而言是最具价值和意义的：一方面是这座桥出现在了合适的位置，以很小的介入重塑了整个村子的空间结构与景观风貌，让村里的通达得到了极大提升，让更多人感受到了幸福；另一方面，或许是因为我们每个人的心中都渴望着一座桥，跨过它以找寻到通向自我的那条路。

建这座桥的同时，相关的河道疏浚与沿线景观、游步道、夜景灯光等建设也同步进行，所以当桥完工的时候，配套设施也几乎同时竣工。它们组合起来后，这里成为玉华村内最具吸引力的景观与活动场所。黄昏时分，灯光亮起，桥上的那几片顶棚晶莹剔透，投射到水面里熠熠生辉，成为村里新的地标。当然，成为地标不是桥的真正目的，真正目的是让人们在此更好地往来、相遇，找到生活的美好。

surface looks absolutely divine. As a new landmark in village, the real purpose of this bridge is to make it possible for people to better connect with each other and find the beauty of life.

桥的建设对村庄空间的改变

The Construction of the Bridge Changes Village Space

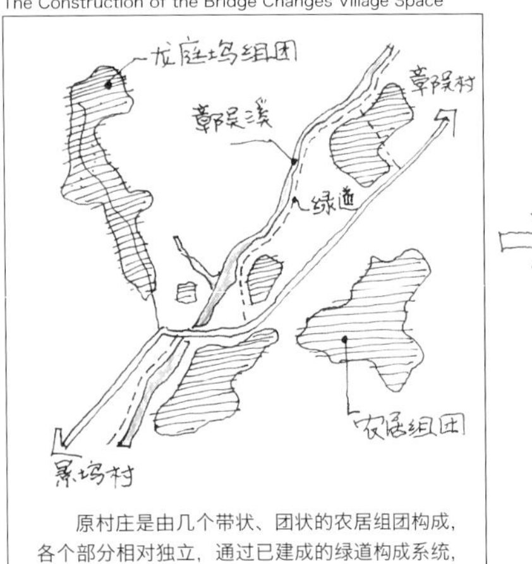

原村庄是由几个带状、团状的农居组团构成，各个部分相对独立，通过已建成的绿道构成系统，而且与日常居住生活的联系不够密切

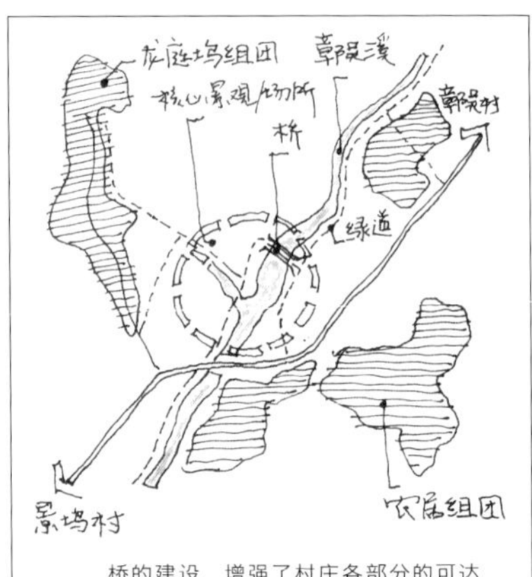

桥的建设，增强了村庄各部分的可达性、环通性，并形成了以几何中心位置的河道水区景观作为村内的核心景观 / 生活场所

平面图

Plan

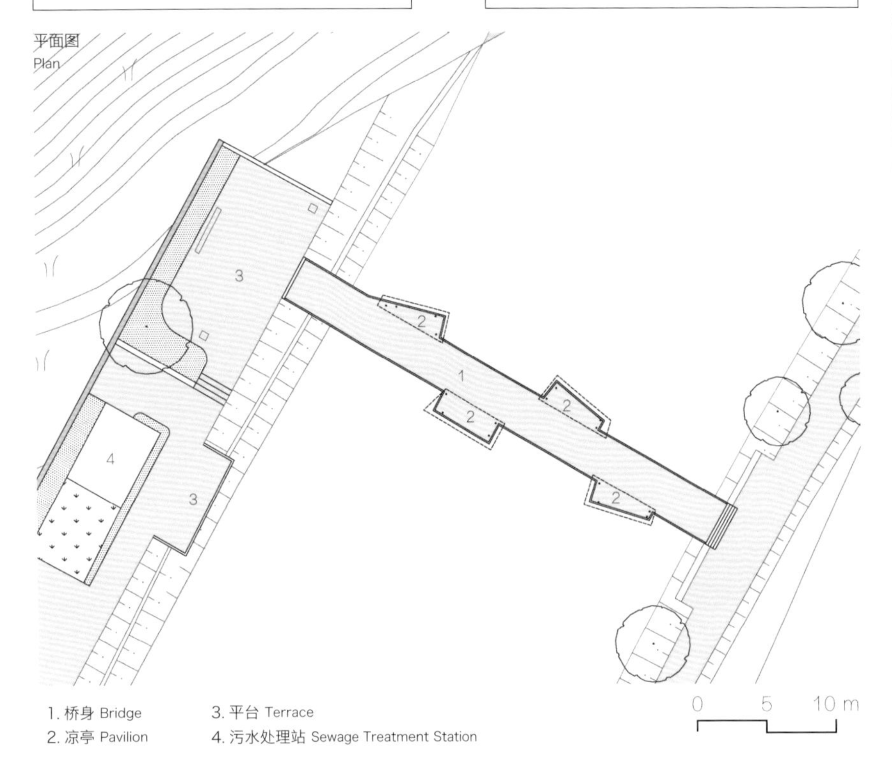

1. 桥身 Bridge
2. 凉亭 Pavilion
3. 平台 Terrace
4. 污水处理站 Sewage Treatment Station

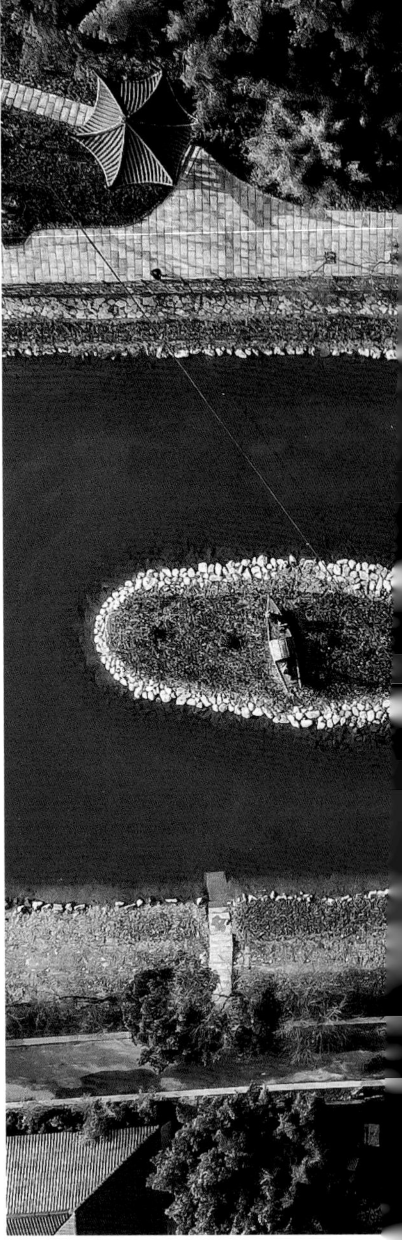

玉华村人行桥

项目在鄣吴镇的位置分布图

The Site of the Project on the Map of Zhangwu Town

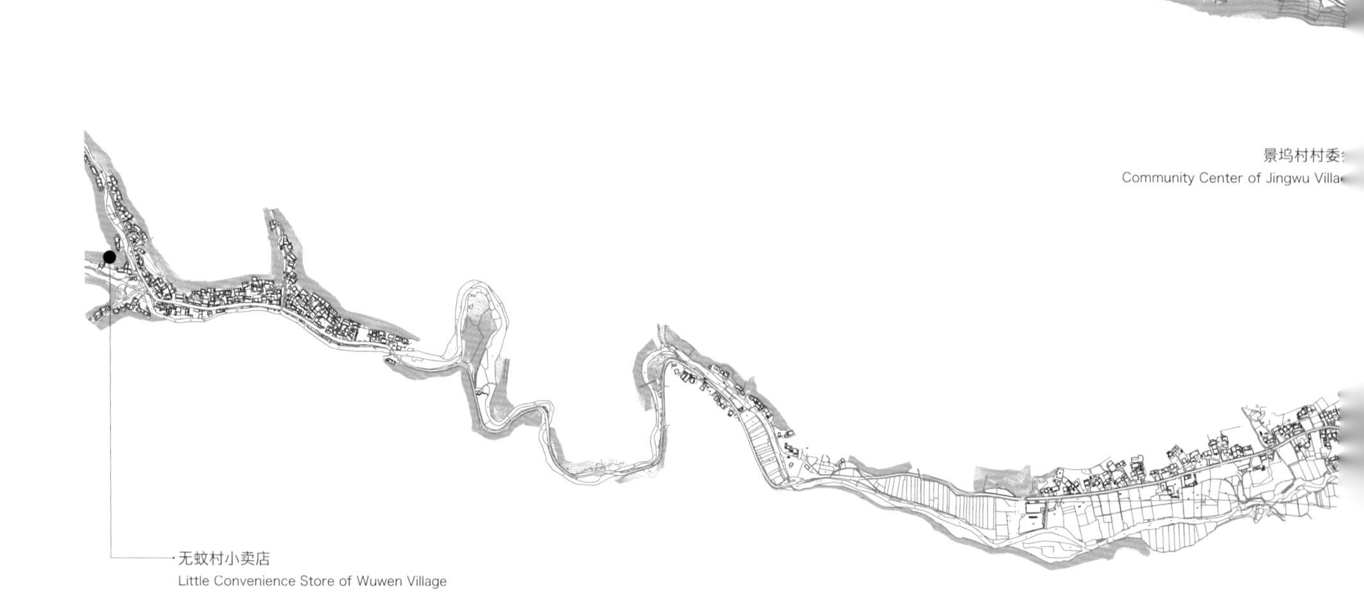

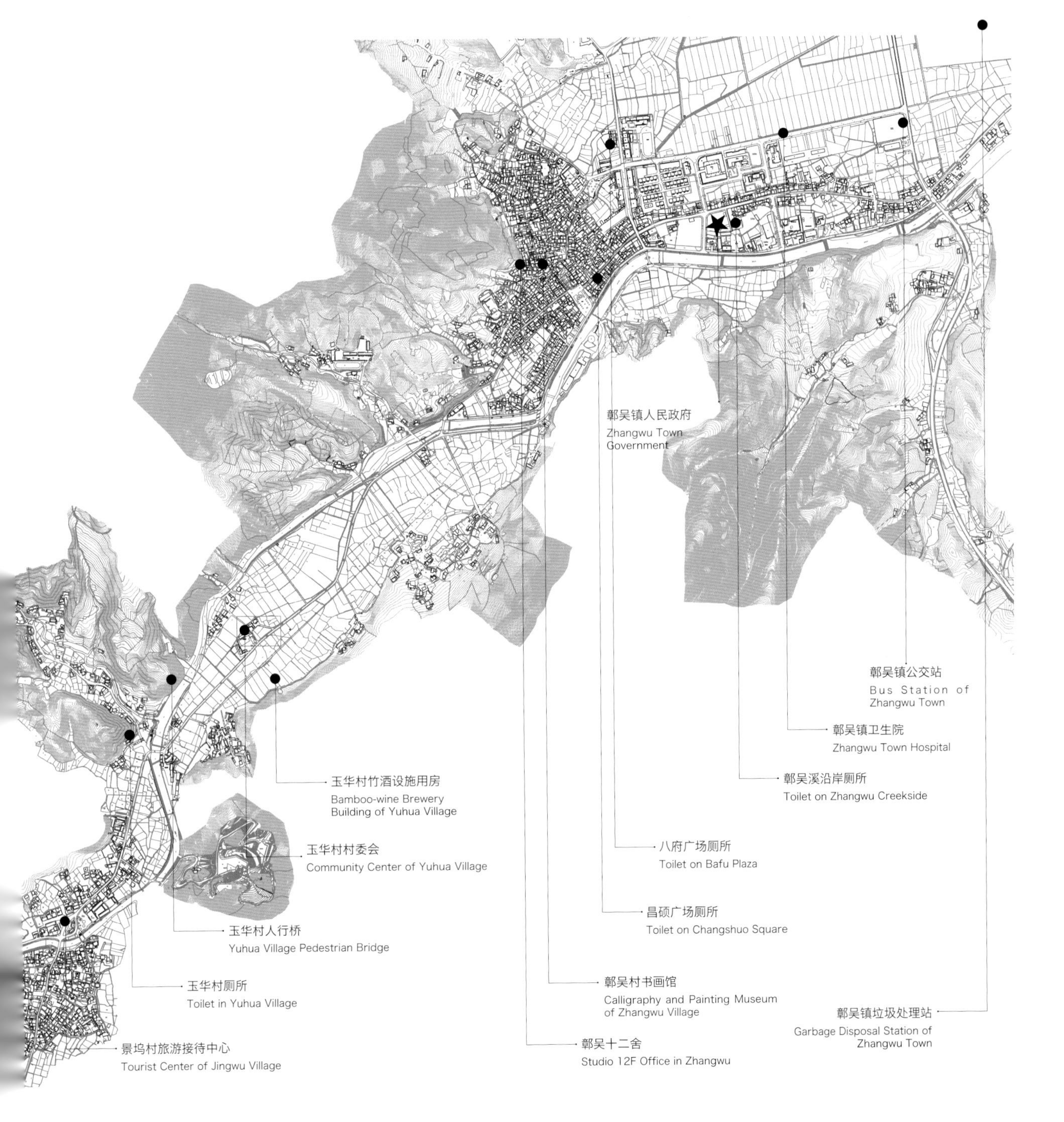

鄣吴镇人民政府
Zhangwu Town Government
鄣吴镇公交站
Bus Station of Zhangwu Town
鄣吴镇卫生院
Zhangwu Town Hospital
鄣吴溪沿岸厕所
Toilet on Zhangwu Creekside
八府广场厕所
Toilet on Bafu Plaza
昌硕广场厕所
Toilet on Changshuo Square
鄣吴村书画馆
Calligraphy and Painting Museum of Zhangwu Village
鄣吴镇垃圾处理站
Garbage Disposal Station of Zhangwu Town
鄣吴十二舍
Studio 12F Office in Zhangwu
玉华村竹酒设施用房
Bamboo-wine Brewery Building of Yuhua Village
玉华村村委会
Community Center of Yuhua Village
玉华村人行桥
Yuhua Village Pedestrian Bridge
玉华村厕所
Toilet in Yuhua Village
景坞村旅游接待中心
Tourist Center of Jingwu Village

项目年鉴
Chronology of Projects

景坞村旅游接待中心
Tourist Center of Jingwu Village

建成时间:
Date of Completion:
2013 年 3 月

鄣吴镇垃圾处理站
Garbage Disposal Station of Zhangwu Town

建成时间:
Date of Completion:
2014 年 10 月

鄣吴村书画馆
Calligraphy and Painting Museum of Zhangwu Village

建成时间:
Date of Completion:
2013 年 6 月

上吴村蔬菜采摘设施用房
Vegetable Planting and Picki Facilities of Shangwu Village

建成时间:
Date of Completion:
2015 年 10 月

无蚊村小卖店
Little Convenience Store of Wuwen Village

建成时间:
Date of Completion:
2013 年 11 月

八府广场厕所
Toilet on Bafu Plaza

建成时间:
Date of Completion:
2015 年 12 月

玉华村厕所
Toilet in Yuhua Village

建成时间:
Date of Completion:
2014 年 4 月

鄣吴镇公交站
Bus Station of Zhangwu To

建成时间:
Date of Completion:
2016 年 2 月

鄣吴镇卫生院
Zhangwu Town Hospital

建成时间:
Date of Completion:
2016 年 5 月

玉华村竹酒设施用房
Bamboo-wine Brewery
Building of Yuhua Village

建成时间:
Date of Completion:
2017 年 9 月

景坞村村委会
Community Center of
Jingwu Village

建成时间:
Date of Completion:
2017 年 1 月

鄣吴溪沿岸厕所
Toilet on Zhangwu Creekside

建成时间:
Date of Completion:
2017 年 12 月

玉华村村委会
Community Center of
Yuhua Village

建成时间:
Date of Completion:
2017 年 3 月

昌硕广场厕所
Toilet on Changshuo Square

建成时间:
Date of Completion:
2017 年 12 月

鄣吴十二舍
Studio 12F Office
in Zhangwu

建成时间:
Date of Completion:
2017 年 7 月

玉华村人行桥
Yuhua Village
Pedestrian Bridge

建成时间:
Date of Completion:
2018 年 12 月

十三楼建筑工作室成员 | Members of Architect Studio 12F